New Frontiers in Regional Science: Asian Perspectives

Volume 25

New Frontiers in Regional Science: Asian Perspectives

This series is a constellation of works by scholars in the field of regional science and in related disciplines specifically focusing on dynamism in Asia.

Asia is the most dynamic part of the world. Japan, Korea, Taiwan, and Singapore experienced rapid and miracle economic growth in the 1970s. Malaysia, Indonesia, and Thailand followed in the 1980s. China, India, and Vietnam are now rising countries in Asia and are even leading the world economy. Due to their rapid economic development and growth, Asian countries continue to face a variety of urgent issues including regional and institutional unbalanced growth, environmental problems, poverty amidst prosperity, an ageing society, the collapse of the bubble economy, and deflation, among others.

Asian countries are diversified as they have their own cultural, historical, and geographical as well as political conditions. Due to this fact, scholars specializing in regional science as an inter- and multi-discipline have taken leading roles in providing mitigating policy proposals based on robust interdisciplinary analysis of multifaceted regional issues and subjects in Asia. This series not only will present unique research results from Asia that are unfamiliar in other parts of the world because of language barriers, but also will publish advanced research results from those regions that have focused on regional and urban issues in Asia from different perspectives.

The series aims to expand the frontiers of regional science through diffusion of intrinsically developed and advanced modern regional science methodologies in Asia and other areas of the world. Readers will be inspired to realize that regional and urban issues in the world are so vast that their established methodologies still have space for development and refinement, and to understand the importance of the interdisciplinary and multidisciplinary approach that is inherent in regional science for analyzing and resolving urgent regional and urban issues in Asia.

Topics under consideration in this series include the theory of social cost and benefit analysis and criteria of public investments, socio-economic vulnerability against disasters, food security and policy, agro-food systems in China, industrial clustering in Asia, comprehensive management of water environment and resources in a river basin, the international trade bloc and food security, migration and labor market in Asia, land policy and local property tax, Information and Communication Technology planning, consumer "shop-around" movements, and regeneration of downtowns, among others.

More information about this series at http://www.springer.com/series/13039

Euijune Kim • Brian H. S. Kim
Editors

Quantitative Regional Economic and Environmental Analysis for Sustainability in Korea

Springer

Editors
Euijune Kim
Department of Agricultural Economics and Rural Development and Research Institute of Agricultural and Life Science
Seoul National University
Seoul, Korea

Brian H. S. Kim
Department of Agricultural Economics and Rural Development, Interdisciplinary Program in Agricultural and Forest Meteology, and Research Institute for Agriculture and Life Sciences
Seoul National University
Seoul, Korea

ISSN 2199-5974 ISSN 2199-5982 (electronic)
New Frontiers in Regional Science: Asian Perspectives
ISBN 978-981-10-0298-4 ISBN 978-981-10-0300-4 (eBook)
DOI 10.1007/978-981-10-0300-4

Library of Congress Control Number: 2016957509

Printed on acid-free paper

This Springer imprint is published by Springer Nature
The registered company is Springer Nature Singapore Pte Ltd.
The registered company address is: 152 Beach Road, #22-06/08 Gateway East, Singapore 189721, Singapore

Preface

Sustainable development, with its dual emphasis on the most recent concerns—development and environment—has different implications for several sections of society, professions, and practitioners in different fields of human endeavor and survival. It requires an extensive application and interdisciplinary method that should form the ingredients of comprehensive approaches to sustainable development. Rigorous theories have been advanced, relatively recently, in response to a growing trend and challenges posed in public policy in economic and environmental development. Economic science recognized nonlinearities in economic growth and environmental conservation and the critical role of environment and ecology in the managing of economic systems. It is therefore necessary to examine more real-life solutions to identified and potential problems that affect the welfare of the human society. However, sustainable development draws much of its significance, influence, and innovation from its very ambiguity. The concrete challenges of sustainable development are at least as heterogeneous and complex as the diversity of human societies and natural ecosystems. A dynamic and evolving idea that can be adapted to fit very different conditions and contexts across space and time allows redefining and reinterpreting the salient features of sustainable development.

New findings of extensive empirical research have posed several new challenges to our basic understanding of how regions develop, change, and grow. This book focuses on lately developed pioneering analytical tools for sustainable development with the application of regional economic and environmental issues in Korea. With a range of case studies, the authors here embraced a series of theoretical models and empirical methods including spatial CCE model, multiregional input–output and econometric analysis, logit model, GIS, contingent valuation method, sample selection model, machine learning technique, stochastic frontier analysis, Markov chain model, and panel analysis. These models and methods are tailored to spatial development and policy issues such as agglomeration, clustering and industrial innovation, human capital and labor market, education and R&D investments, regional economic resilience for unexpected disaster, quarantine system and disease, and environmental degradation. These studies contribute to advocate alternative interpretations and policy guidelines and allow to observe changing patterns and binding constraints of sustainable development transition in Korea.

Seoul, Korea

Euijune Kim
Brian H. S. Kim

Contents

Contributors

Taelim Choi Incheon Development Institute, Incheon, Korea

Ayoung Kim Department of Agricultural Economics, Purdue University, West Lafayette, IN, USA

Brian H. S. Kim Department of Agricultural Economics and Rural Development, Interdisciplinary Program in Agricultural and Forest Meteology, and Research Institute for Agriculture and Life Sciences, Seoul National University, Seoul, Korea

Euijune Kim Department of Agricultural Economics and Rural Development and Research Institute of Agricultural and Life Science, Seoul National University, Seoul, Korea

Jae Hong Kim Department of Public Administration, School of Social Sciences, University of Ulsan, Ulsan, Korea

Changkeun Lee Research Institute of Advanced Materials, Seoul National University, Seoul, Korea

Jaewon Lim School of Public Policy & Leadership, Greenspun College of Urban Affairs, University of Nevada, Las Vegas, NV, USA

Up Lim Department of Urban Planning and Engineering, Yonsei University, Seoul, Korea

Kyung-Min Nam Department of Urban Planning and Design, The University of Hong Kong, Pokfulam Road, Hong Kong

JiYoung Park Department of Urban and Regional Planning, University at Buffalo, The State University of New York, Buffalo, NY, USA

In Kwon Park Department of Urban Administration, University of Seoul, Seoul, Korea

Marc Wolfram Department of Urban Planning and Engineering, Yonsei University, Seoul, Korea

Dong Keun Yoon Department of Urban Planning and Engineering, Yonsei University, Seoul, Korea

Part I
Sustainability and Regional Labor Market

1. The Role of Cities in Sustainability Transitions: New Perspectives for Science and Policy
2. Exploring the Geography of Educational Segregation in Seoul, Korea
3. Labor Market Distortion with Discouraged Worker Effects in Korea

The Role of Cities in Sustainability Transitions: New Perspectives for Science and Policy

Marc Wolfram

Abstract Sustainable development at a global and local scale heavily depends upon the pathways taken by cities in the near future. Within scientific research, this frequently identified "urban challenge" has been recognized and addressed increasingly in urban studies, as well as in transformation studies. However, while both fields clearly overlap and effectively complement each other in this regard, the respective epistemic communities have largely remained separate so far. Therefore, this paper elaborates on the core concepts and approaches that dominate the emerging scientific debate on the role of cities in sustainability transitions. Based on a methodic literature review, it delineates the progressive convergence of the diverse disciplines involved over four major research perspectives. It equally derives key conclusions for future research and policy, highlighting the urgent need to connect the four fields identified, to link socio-technical and social-ecological system (SES) perspectives, to conceive of holistic innovations for developing new planning approaches, and to fully embrace transdisciplinarity by practicing science in society.

Keywords Urban studies • Transformation studies • Sustainability • Epistemology • Transdisciplinarity

1 Introduction

For about a decade, a gradual convergence has taken place between the two interdisciplinary research fields of urban studies and transformation studies. While the former is dedicated to the understanding of cities and their development, the latter explores and explains profound societal and environmental change. With the steady growth of sustainability problems and under the pressure of complex challenges such as climate change and post-fossil energy supply, this convergence of the two research fields increasingly reflects what may turn out to be a necessary

M. Wolfram (✉)
Department of Urban Planning and Engineering, Yonsei University, Seoul, Korea
e-mail: m.wolfram@yonsei.ac.kr

E. Kim, B.H.S. Kim (eds.), *Quantitative Regional Economic and Environmental Analysis for Sustainability in Korea*, New Frontiers in Regional Science: Asian Perspectives 25, DOI 10.1007/978-981-10-0300-4_1

symbiosis: urban studies need to conceive of transformation dynamics, while transformation studies in turn require a better understanding of the role of cities. In order to underpin this basic hypothesis, the essential features of both research fields will first be briefly outlined.

1.1 Understanding Cities: Urban Studies

Constituted only by its shared subject – cities – the scientific field of "urban studies" is a highly interdisciplinary one with fuzzy edges. It comprises *all* scientific perspectives on cities, examining their condition and development across time and space. Thus, there has been a corresponding epistemological and methodological diversity from the outset, since the discursive construction of cities as a subject of science was driven by individual disciplines. This included not only a concern for analysis and interpretation but also for developing and implementing *new forms of intervention*, represented by the subfield of planning (cf. Fainstein and Campbell 2012; Harding and Blokland 2014). It is here that the evolving modes of urban governance and requirements for steering urban development have been discussed extensively (see, e.g., Healey 1992; Albrechts 2004; de Roo and Silva 2010).

Since the 1970s and informed by poststructuralist thinking, urban studies have gradually started to develop a more widely shared ontology, based on the perspective of *relational geography*. Cities became increasingly framed as local nodes within multiple overlapping social, economic, ecological, political, and physical networks, continuously *shaping* and *shaped by* flows of people, matter, and information across scales (Murdoch 2006; Davoudi and Strange 2009).

This poststructuralist shift acknowledged for the crucial role of places in (re) configuring "glocal" power relations and patterns of exploitation (Sassen 1991; Castells 2000; Brenner 2004). It has also been an important catalyst for a broader engagement with the normative concept of *sustainability*, following the 1992 Rio summit. Central epistemological axes in urban studies appeared to resonate particularly well with key tenets of sustainability that demanded holistic thinking and action, including the basic concern for human needs and justice ("inter-/intragenerational equity"); for social, ecological, and economic dynamics ("triple bottom line"); for power and institutions ("good governance"); as well as for place, communities, and culture ("Local Agenda 21") (UN SDSN 2013; Vojnovic 2014). Hence, a broad diversity of boundary disciplines, a relational understanding of space and place, and an orientation at intervention for sustainability are main characteristics of the field that have become important for a growing engagement with transformation studies.

1.2 Understanding System Change: Transformation Studies

Transformation studies are an equally interdisciplinary field, although a considerably smaller and younger one. Its emergence in the early 2000s has been strongly driven by sustainability concerns from the outset. Recognizing the *systemic* character of societal sustainability problems, its subjects are fundamental changes in human-environment (sub)systems that alter ways of thinking (cultures), organizing (structures), and doing (practices). Transitions are understood as coevolutionary processes through which *complex adaptive systems transform*, thus involving multiple actors and action domains. The field is demarcated by a range of conceptual frameworks that theorize on the particular dynamics of transformations, originating from different ontological and epistemological backgrounds (Gunderson and Holling 2002; Folke et al. 2010; Elzen et al. 2004; Markard et al. 2012).

One constitutive strand has been the historical study of large-scale socio-technical systems (STS), unpacking the path-dependent patterns that shape their creation and evolution (Bijker et al. 1987). This has given rise to the *multilevel perspective* (MLP), a heuristic framework that maps interactions between incumbent socio-technical configurations (regimes), alternative solutions in their infancy (niches), and developments in the system environment (landscape) (Geels 2002). Drawing on a variety of related science fields, new frameworks have been conceived to inform policy intervention for sustainability, each engaging progressively with the specificities of the local scale: *transformation management* (TM) builds on the enabling role of governance, foresight, experimentation, and learning in transition processes (Rotmans et al. 2001); *strategic niche management* (SNM) targets the formation, selection, and empowerment of promising niches (Kemp et al. 1998); and *technological innovation systems* (TIS) focus on understanding the actor constellations, institutions, and processes that help or hinder technology breakthrough and mainstreaming (Bergek et al. 2008).

A different constitutive strand forms the study of social-ecological system (SES) and their de- and restabilization, elaborating on the concept of *resilience* (Holling 1973). Similarly, these studies have recognized the need for governance innovations, foresight, knowledge transfers, and learning-by-doing across scales, increasingly linked to urban contexts (Ernstson et al. 2010). Therefore, transformation studies to date offer a range of perspectives on societal change that emphasize different system types (STS, SES), forms of agency, scales, and dynamics of change while reflecting a growing concern for cities as strategic hotspots.

2 Urban Transformation Studies: Epistemological Trajectories

Against this backdrop, the following sections trace the trajectories along which a convergence between the two research fields outlined above has taken place so far and identify common orientations for understanding and shaping urban transformation. To this end, first their mutual engagement is briefly reviewed, highlighting epistemological, empirical, and methodological characteristics (Sect. 2). Next, four principal perspectives are identified that currently appear to dominate and structure the emergent field of urban transformation studies (Sect. 3). Building on these perspectives, conclusions are drawn regarding future requirements in research, policy, and practice (Sect. 4).

The corpus for the literature review has been identified on the basis of a keyword search (Scopus, Web of Science, Google Scholar), concluded in November 2014. Search terms were formed combining core terminology from both fields ("system transition," "system transformation," urban, city, region, space, spatial, place, scale). Pertinent references were selected by reviewing abstracts and conclusions, thereby excluding divergent understandings (e.g., "urban transition" as *urbanization*). Further relevant sources have then been identified successively through the reference lists included, finally retaining a total of 115 references for analysis (93 journal articles, 18 books, 4 book chapters). These have each been reviewed independently by at least two different researchers to specify five basic characteristics: (1) normative position (sustainability), (2) interdisciplinarity, (3) main concepts and theories used, (4) methodology, and (5) empirical sources. On this basis, further subcategories could be established for the second and third review, thus creating a more differentiated typology from the bottom up (Table 1).

2.1 Transformation Studies and the City

The epistemological concern of transformation studies for system dynamics has increasingly also triggered research dealing with the role of cities. Most references analyzed adopt system transformation theory frameworks and concepts as a heuristic to explore patterns and dynamics of urban change. But also a normative use to develop new forms and methods for steering and intervention is frequent, given the orientation at sustainability.

The most widely used conceptual frame is the "multilevel perspective" (MLP), followed by "transition management" (TM), "resilience," and "innovation systems" (Table 1). "Coevolution" and "social innovation" concepts are hardly employed independently from these (Roggema et al. 2012; see, e.g., Mader 2013), with the combination between "social innovation" and "strategic niche management" (SNM) forming a more persevering pattern. Also, very few researchers draw on both the MLP *and* "resilience" (see, e.g., Newton 2008; McCormick et al. 2013), which reflects a clear divide between the respective epistemic communities rooted in either STS or SES scholarship. Thus, apart from the prevailing use of the MLP, two related trajectories are informed by TM or SNM/"social innovation" and two more independent ones build on "innovation systems" or "resilience" theory.

The *MLP* maps out how niche-regime interactions affect the creation and unfolding of pathways for socio-technical transitions that lead to new system configurations. By adopting the MLP, new basic questions have thus been raised

Table 1 Bottom-up categorization of references and total incidence ($n = 115$)

Category	Subcategory	No.	%
Normative	Sustainability	100	87%
Interdisciplinarity	Urban transformation studies (integrated approaches)	31	27%
	Transformation theory in urban studies	31	27%
	Urban subjects in transformation theory	20	17%
	Urban theory in transformation studies	19	17%
	Transformation subjects in urban theory	14	12%
Urban theory/concepts	Urban Governance [1]	75	65%
	Urban Planning (incl. regional planning)	65	57%
	Urban Geography (incl. economic and political geography)	40	35%
	Urban Ecology (incl. political and industrial ecology)	31	27%
	Urban Sociology [2]	18	16%
	Urban Design (incl. architecture, building engineering)	15	13%
	Other (urban theory) [3]	14	12%
Transformation theory/concepts	Multi-Level Perspective (MLP)	52	45%
	Transition Management (TM)	36	31%
	SES resilience [4]	27	23%
	Co-evolution	21	18%
	Innovation Systems (incl. regional-, local- and technological-)	20	17%
	Social innovation (incl. social practice theory, social movement theory)	19	17%
	Other (transition theory) [5]	11	10%
	Strategic Niche Management (SNM)	7	6%
Methods	Deduction / Hermeneutics	82	71%
	Case study	75	65%
	Transdisciplinarity	7	6%
	Data mining	5	4%
	Modelling	2	2%
Empiricism	City	78	68%
	Region	49	43%
	Building / neighborhood	33	29%
	National	19	17%
	Networks (of cities or initiatives)	13	11%
	Practice	88	77%
	Policy	62	54%
	None	13	11%
Epistemology	A: Transforming urban metabolisms and political ecologies	58	50%
	B: Configuring urban innovation systems for green economies	36	31%
	C: Building adaptive urban communities and ecosystems	29	25%
	D: Empowering urban grassroots niches and social innovation	25	22%
TOTAL	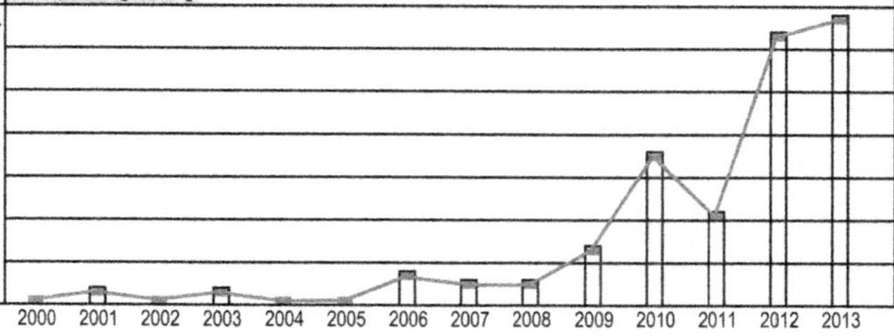	35 30 25 20 15 10 5	p.a.

1) Incl. multi-level governance, neo-institutionalism, regulation theory, discourse theory.
2) Incl. organization and knowledge sociology, anthropology, culture theory.
3) Incl. policy analysis, economics, information society studies, transport studies, rural studies, environmental justice.
4) Incl. adaptive and transformative capacity, adaptive renewal cycle, panarchy.
5) Incl. actor network theory, sustainability assessment, market transformation, complexity theory, spatial transition, sustainable consumption and production, ecological restructuring, social construction of technology.

for urban policy and planning that address the role of cities as purposeful *actors* in socio-technical transitions and their possible influence on (national) regime transformation and/or as *seedbeds* for local innovation niches (Geels 2011; Raven et al. 2012), especially with a view to urban infrastructures (Bulkeley et al. 2011). But also cities themselves have been interpreted as *urban regimes*, configured through strategic work by incumbent urban actors (Quitzau et al. 2013). However, the MLP has only rarely served to directly derive new approaches for urban policy and planning that address niche-regime constellations (see, e.g., de Graaf and van der Brugge 2010), but mostly required a combination with other transformation and/or urban theory to meaningfully address intervention options (cf. Burch et al. 2014).

Here, especially *TM* has been helpful as a heuristic to examine the characteristics of urban governance and planning processes. Based on its postulations concerning actor types (frontrunners, border crossers, incumbents), interaction forms (transition arena), and activities (orientating, agenda setting, activating, reflecting) (Rotmans 2006; Loorbach and Rotmans 2010), TM has been largely employed for empirical assessment and/or development of urban policy guidance. Focused on single domains such as water management (Brown et al. 2013) or information infrastructures (Wolfram and Vogel 2012), or regarding broader development strategies such as waterfront regeneration (Frantzeskaki et al. 2013), this has provided deeper insights concerning the role of agency and leadership, as well as pilots and experiments for enabling transformative governance and social learning in urban contexts. It has equally informed the conception of action research in "urban transition labs," i.e., transdisciplinary interaction spaces that complement existing governance arenas (Nevens et al. 2012). Moreover, the specific design of foresight processes within TM has generated suggestions for modifying urban planning methods (Wiek et al. 2003; Eames and Egmose 2011).

Other studies have invoked *SNM* as a conceptual reference in order to "zoom in" on the requirements of local niches and their relations to socio-technical regimes. While this has largely confirmed the importance of general success factors identified in the SNM literature such as shared stakeholder expectations, enabling actor networks and experiential learning (Bai et al. 2010; Schreuer et al. 2010), it has also illustrated the need for a better understanding of locally *embedded* niches. Some scholars have therefore linked SNM with social innovation theories in order to also trace the implications of practical know-how, physical activities, and cultural meanings for the transformative impact of community initiatives and "grassroots niches" (Davies 2012; Seyfang and Haxeltine 2012; Smith and Seyfang 2013) – yet often without a clear differentiation of their urban and spatial contexts. By contrast, others have strongly underlined the need to acknowledge for the *place-specific* constitution of niches and related options for strategic urban planning (Quitzau et al. 2012) or a less antagonistic but more relational understanding of locally shaped niche-regime configurations (Maassen 2012). Furthermore, the focus on urban niches has also led to recognize the necessity to develop new approaches to public participation in urban planning with a view to enable civil society and private sector actors to effectively contribute to urban transformations (Aylett 2013).

A different direction has been pursued by those drawing on (technological) *innovation systems* studies and their concern for the institutions and actor networks that shape the creation, adoption, and diffusion of new technologies (cf. Hekkert and Negro 2009). As recognized by earlier approaches to managing the transformation of local socioeconomic systems (Wiek et al. 2003), embedded actor strategies and institutional structures can become vital factors for the breakthrough of "green" technologies, industries, and markets (Coenen et al. 2012). Empirical studies illustrate this for specific technologies (Carvalho et al. 2012; Dewald and Truffer 2012) or clusters of eco-innovation (Cooke 2010; Cooke 2011; Lahlou 2011; McCauley and Stephens 2012) while simultaneously highlighting the multilevel character of the processes observed. This underlines the unique position of cities as the places that connect consumers, producers, and policy, shaping global consumption patterns through urban lifestyles (Reusswig 2010).

Last but not least, *resilience theory* has generated another rich strand of research addressing urban sustainability transitions, yet starting from an understanding of cities as social-ecological systems to identify vulnerabilities, unsustainable performances, and dynamics of change. Based on the adaptive renewal cycle and the concept of panarchy, there is a very similar concern for governance innovations, experimentation, and social learning (Ernstson et al. 2010; Folke et al. 2010). This is reflected in a comparable spectrum of research interests with a view to interpret overall urban transformation dynamics and identify options for steering (Pincetl 2012; Wilson 2012; Pickett et al. 2013; Cole et al. 2013), develop orientation and practical guidance for urban planning and design (Desouza and Flanery 2013; Jabareen 2013; Lu and Stead 2013) and related foresight (Van der Voorn 2012), and explain the emergence and impact of local innovations (Boyd and Ghosh 2013), as well as related lifestyle changes (Peters et al. 2012).

2.2 *Urban Studies and Transformative Change*

Inversely, a number of researchers have used urban theory to explore the role of cities and regions in relation to transformations thus engaging critically with concepts from both STS studies and resilience theory. This has allowed to not only substantiate a call for better recognizing the crucial role of space in mainstream conceptions of socio-technical transitions. More importantly, it has enabled a differentiated account for the specific dynamics resulting for and from cities with a view to sustainability transitions. Four main strands need to be distinguished here by the nature of their concerns and the theoretical references used, respectively rooted in *economic geography*, *political ecology*, *sociology and anthropology*, or *planning studies*.

Research informed by *economic geography* has been particularly instrumental to acknowledge for the spatial implications of transitions and to also provide adequate concepts to capture these. Following earlier calls for a "geography of sustainability transitions" (Smith et al. 2010), the endeavor has here been to foreground the role of place and scale in transitions, thereby enhancing the utility of the MLP. Drawing on relational geography, regime and niche actors have thus been framed within cross-scale spatial and institutional contexts that produce enabling and constraining effects for socio-technical transitions in terms of identity, legitimacy, actor coalitions, and resources. Consequently, the impacts of embeddedness and territorial power relations on creating multi-scalar trajectories and patterns of uneven distribution have been disclosed (Coenen et al. 2012; Truffer and Coenen 2012), especially looking at energy systems (Essletzbichler 2012; Bridge et al. 2013). Regarding cities, this has served to illustrate their ambivalent role in shaping transitions both as places of innovation and as a local manifestation of multi-scalar socio-technical regimes.

A second central motive for using urban theory in transformation studies has been the *political ecology* of resource flows underpinning urban development. Through the lens of urban governance and regime theory, new insights have been obtained into the ongoing reconfiguration of the networked infrastructures that mediate those flows (Guy et al. 2001; Monstadt 2009). Focused on the public and private key stakeholders, their interactions, and the institutional shifts they create, this perspective has illustrated the multilevel and multi-sectoral character of urban socio-technical change (Späth and Rohracher 2012; Uyarra and Gee 2012; Burch et al. 2013), but also the crucial role of strategic local planning processes and new intermediaries (Truffer et al. 2010; Hodson and Marvin 2010; Guy et al. 2011; Bulkeley et al. 2011, 2012; Hamann and April 2012). It has equally underlined how urban experiments and civil society participation contribute to articulate new system configurations in concrete settings (Castán Broto and Bulkeley 2013; Rydin et al. 2013). In order to orient transitions toward sustainability and avoid new elitist forms of steering (Khan 2013), different new requirements have been identified for urban governance and planning (Domènech and Saurí 2010; Young 2010; Scerri and Holden 2013). Especially participatory foresight and novel forms of intermediation turn out to be critical elements in transition processes grounded in urban contexts with a view to their contribution to create shared visions, operational capacity to act, and opportunities for social learning (Späth and Rohracher 2010; Hodson et al. 2013).

Third, increasing attention has been paid to the influence of social practices, communities, and grassroots initiatives on socio-technical transitions, drawing on *sociology and anthropology*. While recognizing the steering attempts of urban regime actors, this perspective acknowledges especially for the time- and space-specific constituents of everyday practices as equally basic conditions for system innovations (Shove and Walker 2010). Hence, differences between places in terms of discourses, cultural frames, and identity result to be critical factors for transformative governance that require attention through foresight, community participation, and empowerment approaches (Mulugetta et al. 2010; Cooke and Rehfeld 2011; Marsden 2012; Späth 2012). The case of the UK Transition Town movement and its diffusion has received much attention from this perspective, giving rise to critical questions regarding transition visions, politics, and culture (Brown et al. 2012; Mason and Whitehead 2012; Neal 2013). These studies clearly recognize that cities provide far better opportunities for scaling up the impacts of grassroots initiatives than the villages and small towns that currently prevail in this particular movement. Especially the capacity to empower communities and to draw on translocal and cross-scale networks appears to be a crucial asset of cities (Taylor 2012; North and Longhurst 2013).

Finally, *planning studies* have increasingly turned toward urban transformations as well, both conceptually and empirically. Starting from earlier engagements with complexity theory and its lessons for planning in terms of handling uncertainty, thresholds, and emergence (Innes and Booher 1999; de Roo and Silva 2010), requirements for planning processes to explicitly address transitions have gradually become further specified. This has underlined the pertinence of the theoretical debates on collaborative, adaptive, and/or strategic urban planning, especially regarding their emphasis on participation, knowledge co-creation, long-term foresight, experimentation, and flexibility (Healey 2007; Truffer et al. 2010; Rauws and de Roo 2011). While some authors have sought to substantiate their conceptual considerations through the strategies and measures recognized in current planning practice (Portney 2009; Hagelskjær Lauridsen and Stissing Jensen 2013), others have discussed conceptual ambiguities when applying transformation theory to cities. This concerns especially the constitution of cities out of multiple coalescing subsystems, both socio-technical and social-ecological, that require to conceive of "multi-regimes" and to develop different strategies for managing place-based niches in a highly inert built environment (Næss and Vogel 2012; Quitzau et al. 2012).

2.3 Methods, Empiricism, and Transdisciplinarity

Regarding the research designs used across all references analyzed, it is first of all the high proportion (1/3) of purely deductive and/or hermeneutic approaches that calls the attention, apparently reflecting lively and ongoing theory development in this field. The empirical work is almost exclusively based on qualitative case studies, with only a few methodological exceptions (surveys, modeling, data mining). Although the majority of these case studies focus on the scale of the city, there are also a number of cross-scale studies that address either relations between the urban/regional and urban/national scales or relations within cities and their subscales of districts, blocks, or buildings. To speak of "urban" transformations is therefore by no means an attempt of spatial delimitation, but rather a necessary focus in the perspective of relational geography, which recognizes the particular importance of cities.

However, there are a number of significant empirical gaps emerging. The case studies invariably deal with *individual* cities – comparative research dealing with several cities has hardly been undertaken, although this would be particularly informative (esp. if realized within the same nation-state to control context variables). Also studies on translocal relations of cities and the role of city networks have been rare so far. Moreover, regarding the geographical location of the cities studied, the empirical basis appears to be largely concentrated in Western Europe, thus (implicitly) assuming specific political, cultural, and socioeconomic conditions. Likewise, despite all interdisciplinarity, there is still a lack of genuine contributions from key fields in urban studies such as planning, engineering, political science, economics, or sociology. These disciplines could however contribute to examine important facets of urban transformations in more depth (or have already done so – yet without invoking transformation theory).

Above all, the proportion of *transdisciplinary* research – i.e., interdisciplinary studies defined and realized together by science and society stakeholders – is surprisingly low. Although the crucial importance of transdisciplinarity for collective knowledge production and learning processes in transitions has been repeatedly emphasized and illustrated (Wiek et al. 2006; Scholz 2011; Mieg and Töpfer 2013; North 2013), and although the urban context provides ideal conditions for transdisciplinary research, practical implementation falls short of meeting this requirement. The case of an international "network for sustainable urban development" recently formed by research institutes and cities represents a pioneering exception here (Childers et al. 2014), but also points to the continued lack of adequate concern in mainstream policy and research.

3 Mapping Perspectives on Urban Transformations

Drawing on the trajectories outlined above, this section provides a more foresighted reading of the references analyzed with a view to inform a future agenda for research, policy, and practice. It highlights four research perspectives that have so far dominated the debate and are therefore also well substantiated both empirically and conceptually. These perspectives are characterized by their emphasis on *distinct drivers of change* (cf. McCormick et al. 2013; Mieg 2013) and their role in shaping urban transformations which implies a particular epistemology (questions, subject, theory, methods). By focusing on a consistent combination of drivers in terms of *agency* (public sector, civil society, private sector) and *system dynamics* (social, economic, ecologic), each perspective thus anticipates a distinctive *urban transformation pathway*, identifying pertinent action domains, stakeholders, and interactions that in turn require corresponding forms of intervention. Without claiming comprehensive coverage or unique attribution of references, the following four salient research perspectives and related pathways have been identified.

3.1 *Transforming Urban Metabolisms and Political Ecologies*

This perspective highlights the strategic responses that powerful urban actors create to the challenge of a shifting political ecology and economy of cities in times of global resource scarcity and climate change. It recognizes that especially local governments and major infrastructure and technology providers increasingly engage in novel forms of place-specific interaction and socio-technical experimentation concerning urban energy, water, waste, or transport. To secure long-term access to vital resources for continued economic growth and safeguarding local assets and living standards, these actors form new alliances that aim to significantly reduce a city's carbon and ecological footprints. New technologies, services, and usages are therefore trialed in urban settings, involving various stakeholders – from industry to NGOs and citizens. Studies adopting this perspective are also wary of scalar relations and multilevel interactions in this, with a view to state institutions, resource markets, or (inter)national companies, and often account for the role of intermediaries and their capability to facilitate change by supporting new visions, discourses, networks, and coalitions. Particular attention is paid here to emerging deficits in terms of legitimacy, accountability, and openness. Hence, this research strongly focuses on STS that condition the urban metabolism and its changing (multilevel) governance. Drawing on the MLP, cities represent complex socio-technical *niches* that can challenge large-scale resource regimes, but also place-based *urban regimes* for small-scale experiments. Together, the forms of agency involved in both constellations are deemed to enable or constrain wider sustainability transition dynamics. This also suggests particular forms of intervention, like strategic networking, intermediation, and/or participatory foresight in order to influence or counterbalance the direction and speed of these processes.

3.2 *Configuring Urban Innovation Systems for Green Economies*

While the central motif of the key actors in this perspective is similar to the previous one (i.e., adjustment to global environmental change in order to stay competitive), "transitions" primarily concern production and consumption patterns here, not (only) infrastructures. Yet, cities are equally vital for this: the focus is on private companies, consumers, and markets for high-/low-carbon products and the place-specific requirements, strategies, and networks for "greening" the related parts of the economy. Actor constellations are recognized that bring together government agencies, industry, SMEs, and academic institutions, jointly initiating and driving innovation processes that improve their competitiveness, while also contributing to reduce the resource intensity of certain products and services. In this, knowledge transfers and innovation activities are conditioned by the formal and informal networks among these actors and the associated formation of shared value systems and cooperation cultures. However, issues of legitimacy or accountability are not necessarily a particular concern here. In this perspective, change for sustainability thus takes place through innovation systems for selected markets and socio-technical practices anchored in cities. This points toward a proactive pursuit of local "public-private-research" cooperations facilitated through certain types of intermediaries (e.g., economic promotion agencies, cluster managers), as well as specific forms of experimentation and open innovation (e.g., Living Labs).

3.3 *Building Adaptive Communities and Ecosystems*

Climate change, resource scarcity, and biodiversity loss form the combined drivers in this perspective, yet especially with a view to the resulting vulnerabilities of cities. Diverse urban stakeholders respond to this challenge, aiming to create a dynamic social-ecological system balance while controlling the local impacts of global environmental change. System relations and contexts considered are thus defined essentially through ecosystem services. Therefore, water supply and catchment areas, building material imports and exports, food provision and agriculture, or green infrastructures and their different functions (carbon sink, water resorption, species protection, shading, recreational space, etc.) are important starting points for future pathways. In this, also a broad variety of locations and typologies needs to be considered (e.g., for green infrastructures: riverbanks, parks, gardens, brownfields, roofs, facades, streets, squares). Correspondingly, the social-ecological interactions and actor constellations are rather diverse but highly inclusive, ranging from the vegetable garden at the scale of the block to material recycling and urban mining in metropolitan areas. Pertinent communities may thus include citizens (as dwellers, owners, users) and civil society groups (local), government agencies as well as private companies, and research institutions. New system configurations can be enabled through fostering self-organization capabilities and creating diverse and redundant solutions. Thus, participation, knowledge, co-production, learning by doing, and adaptive governance become necessary cornerstones of urban policy making and planning.

3.4 *Empowering and Harnessing Urban Grassroots Niches*

In this perspective, change for sustainability is driven by heterogeneous approaches and initiatives of civil society actors in cities. Global environmental change plays an equally crucial role, but responses are rather justified ethically and also need to be seen in relation to other individual and group-specific needs (e.g., employment, housing, mobility) and motives (e.g., identity, self-achievement, recognition, cohesion, solidarity). Correspondingly, there is a wide range of activity fields addressed, including food, education, health, and also green space or renewable energy. This implies that characteristics of urban structure and design such as density, typology, functional mix, and accessibility are of considerable importance since they have a direct or indirect bearing upon stakeholders' means and ends. On the other hand, this interweaving with the built environment also conditions an integrated handling of socio-technical and social-ecological problem dimensions (e.g., as in street rehabilitation or residential- and roof gardens). The focus is on the ability and opportunity of the respective initiatives to promote and scale their innovative practices, both through replication and through translation into policies and regulation, or new markets. The transformative potential of such urban niches is seen to depend on the local institutional cultures and practices, but also translocal relations (peer to peer). Cities may thus appear as innovation incubators, actively empowering and promoting grassroots initiatives and networks, or as regimes that offer structural resistance, and most likely are both at once.

4 Conclusions and Outlook

Based on a methodic literature review, this chapter has discussed why and how the two interdisciplinary fields of urban studies and transformation studies are converging toward research into the urban dimension of complex system changes for sustainability. It has described the emergence of a certain range of epistemological trajectories that have required and fostered an increasing interconnection between both fields. These underline the need to conceive of and study *urban sustainability transformations* with a view to both the characteristic immediacy, imbrication, and variety of innovation dynamics in cities and their strong implications for global (un) sustainability on a fast urbanizing planet. However, it has also recognized a predominant orientation of research at four salient pathways and the corresponding combinations of agency and system dynamics. Based on these findings, and considering earlier roadmap suggestions for transition studies (STRN 2010), future action in science, policy, and practice dealing with urban sustainability transformations would strongly benefit from addressing the limitations and gaps of the pattern identified. Hence, the following issues should inform a shared future agenda in order to move from convergence to synergy and to focus limited resources on high-impact challenges:

1. Studies that have engaged with urban transformations have so far largely drawn on *selected* theoretical constructs to conceive of and explain change. This necessarily implied a more fragmented account for the urban and its role in transformations. While these perspectives remain valid and useful, much could be gained from conceptualizing and exploring interdependencies between the different change dynamics they address, without aspiring to create a "great unified theory." For this, relational geography and (multilevel) governance theory provide shared frameworks that enable a crossover, including between

the various underlying ontologies (cf. Geels 2010). Such a multifaceted theoretical framework could help to create a more adequate understanding of how "places produce transitions and transitions produce places" (STRN 2010, 18). Particular attention should then be paid to emerging synergies and conflicts between pathways and their respective drivers, between orientations at resilience or transformation (cf. Smith and Stirling 2010), and between phasing out *old* and building up *new* regimes (Loorbach 2014). It would equally allow to identify *new tipping points* that effectively couple various innovation dynamics.

2. Knitting the above theoretical framework necessarily entails a shift in terms of the subjects and questions dealt with. As recognized by various authors, looking at cities requires to acknowledge for "multi-regime" configurations that interconnect various STS. However, cities can equally be depicted as a set of coalescing SES that govern diverse stocks, flows, and ecosystem services. Therefore, it becomes crucial to empirically explore how institutions, discourses, actor constellations, and practices avoid or embrace this "hybrid" reality of cities as *social-ecological-technological systems* (SETS) (cf. McGinnis and Ostrom 2014) – and with what implications for transformations. More emphasis needs to be put on the role of *urban place* as a key entity in this, since it is through particular physical landscapes, built environments, identities, and sociocultural practices that such hybrid configurations become manifest in cities. Across the spectrum of epistemological trajectories identified, this raises new questions about how multiple transformation dynamics play out in different places, accounting for their local constitution, as well as translocal and scalar relations.
3. Having recognized the critical role all pathways attribute to agency, leadership, and intermediation, particular efforts need to be undertaken with a view to develop suitable urban approaches for intervention to help initiate, accelerate, and navigate sustainability transformations. Transition management and its local adaptations provide only first orientations here. In addition, capacity building and civil society empowerment form equally important approaches, especially considering the diversity of context conditions and starting points of cities from across the globe. Most importantly, the gap toward urban planning and policy making must be closed in theory and practice. Instruments and techniques applied in this domain (e.g., strategic planning, SEA, foresight, community participation, urban regeneration) offer considerable potential regarding their integrative, governance, and experimental functions, but would require more tailored modifications. Therefore, transcending the available approaches to develop new forms of transdisciplinary "up-down" governance, intermediation, and institutional entrepreneurship in cities is a necessity that would also help in facing the legitimacy challenge of transformation-oriented intervention.
4. The current empirical basis and range of research methods require strategic extensions in various directions. Identifying lessons and patterns regarding the multitude of individual case studies carried out so far seems an immediate requisite. Correspondingly, more emphasis needs to be put on *comparative research*, including both qualitative case study and quantitative analysis of

larger urban data sets, while also looking at failed or locked-in transition pathways. This should be included to widen and balance the empirical basis toward the global South and East, enabling an exploration of the influence of key context variables, but also interconnections between cities and/or regions. Last but not least, the role of transdisciplinarity needs to be strengthened substantially, using especially research policy and programs as a lever to codevelop and mainstream (new) methods for targeted urban interaction between science and society.

References

Albrechts, L. 2004. Strategic (spatial) planning reexamined. *Environment and Planning B: Planning and Design* 31: 743–758. doi:10.1068/b3065.

Aylett, A. 2013. Networked urban climate governance: Neighborhood-scale residential solar energy systems and the example of Solarize Portland. *Environment and Planning C: Government and Policy* 31: 858–875. doi:10.1068/c11304.

Bai, X., B. Roberts, and J. Chen. 2010. Urban sustainability experiments in Asia: Patterns and pathways. *Environmental Science & Policy* 13: 312–325. doi:10.1016/j.envsci.2010.03.011.

Bergek A., Jacobsson S., Carlsson B., et al. 2008. Analyzing the functional dynamics of technological innovation systems: A scheme of analysis. Research Policy 37:407–429. doi: 10.1016/j.respol.2007.12.003.

Bijker, W., T.P. Hughes, and T. Pinch (eds.). 1987. *The social construction of technological systems: New directions in the sociology and history of technology*. Cambridge, MA: MIT Press.

Boyd, E., and A. Ghosh. 2013. Innovations for enabling urban climate governance: Evidence from Mumbai. *Environment and Planning C: Government and Policy* 31: 926–945. doi:10.1068/c12172.

Brenner, N. 2004. *New state spaces: Urban governance and the rescaling of statehood*. Oxford: Oxford University Press.

Bridge, G., S. Bouzarovski, M. Bradshaw, and N. Eyre. 2013. Geographies of energy transition: Space, place and the low-carbon economy. *Energy Policy* 53: 331–340. doi:10.1016/j.enpol.2012.10.066.

Brown, G., P. Kraftl, J. Pickerill, and C. Upton. 2012. Holding the future together: Towards a theorisation of the spaces and times of transition. *Environment and Planning A* 44: 1607–1623. doi:10.1068/a44608.

Brown, R.R., M.A. Farrelly, and D.A. Loorbach. 2013. Actors working the institutions in sustainability transitions: The case of Melbourne's stormwater management. *Global Environmental Change* 23(4): 701–718. doi:10.1016/j.gloenvcha.2013.02.013.

Bulkeley, H., V. Castán Broto, M. Hodson, and S. Marvin (eds.). 2011. *Cities and low carbon transitions*. New York: Routledge.

Bulkeley, H., V. Castán Broto, and G. Edwards. 2012. Bringing climate change to the city: Towards low carbon urbanism? *Local Environment* 17: 545–551. doi:10.1080/13549839.2012.681464.

Burch, S., H. Schroeder, S. Rayner, and J. Wilson. 2013. Novel multisector networks and entrepreneurship: The role of small businesses in the multilevel governance of climate change. *Environment and Planning C: Government and Policy* 31: 822–840. doi:10.1068/c1206.

Burch, S., A. Shaw, A. Dale, and J. Robinson. 2014. Triggering transformative change: A development path approach to climate change response in communities. *Climate Policy* 14(4): 1–21. doi:10.1080/14693062.2014.876342.

Carvalho, L., G. Mingardo, and J. Van Haaren. 2012. Green urban transport policies and cleantech innovations: Evidence from Curitiba, Göteborg and Hamburg. *European Planning Studies* 20: 375–396. doi:10.1080/09654313.2012.651801.

Castán Broto, V., and H. Bulkeley. 2013. A survey of urban climate change experiments in 100 cities. *Global Environmental Change* 23: 92–102. doi:10.1016/j.gloenvcha.2012.07.005.

Castells, M. 2000. *The rise of the network society*. Oxford: Blackwell Publishers.

Childers D.L., Pickett S.T.A., Grove J.M., et al. 2014. Advancing urban sustainability theory and action: Challenges and opportunities. Landscape and Urban Planning 125:320–328. doi: 10.1016/j.landurbplan.2014.01.022.

Coenen, L., P. Benneworth, and B. Truffer. 2012. Toward a spatial perspective on sustainability transitions. *Research Policy* 41: 968–979. doi:10.1016/j.respol.2012.02.014.

Cole, R.J., A. Oliver, and J. Robinson. 2013. Regenerative design, socio-ecological systems and co-evolution. *Building Research and Information* 41: 237–247. doi:10.1080/09613218.2013.747130.

Cooke, P. 2010. Regional innovation systems: Development opportunities from the "green turn.". *Technology Analysis & Strategic Management* 22: 831–844. doi:10.1080/09537325.2010.511156.

Cooke, P. 2011. Transition regions: Regional–national eco-innovation systems and strategies. *Progress in Planning* 76: 105–146. doi:10.1016/j.progress.2011.08.002.

Cooke, P., and D. Rehfeld. 2011. Path dependence and new paths in regional evolution: In search of the role of culture. *European Planning Studies* 19: 1909–1929. doi:10.1080/09654313.2011.618685.

Davies, A. 2012. *Enterprising communities: Grassroots sustainability innovations*, Advances in ecopolitics 9. Bingley: Emerald.

Davoudi, S., and I. Strange (eds.). 2009. *Conceptions of space and place in strategic spatial planning*. New York: Routledge.

de Graaf, R., and R. van der Brugge. 2010. Transforming water infrastructure by linking water management and urban renewal in Rotterdam. *Technological Forecasting and Social Change* 77: 1282–1291.

de Roo, G., and E. A. Silva. 2010. *A planner's encounter with complexity*. Farnham (UK)/ Burlington (US): Ashgate.

Desouza, K.C., and T.H. Flanery. 2013. Designing, planning, and managing resilient cities: A conceptual framework. *Cities* 35: 89–99. doi:10.1016/j.cities.2013.06.003.

Dewald, U., and B. Truffer. 2012. The local sources of market formation: Explaining regional growth differentials in German photovoltaic markets. *European Planning Studies* 20: 397–420. doi:10.1080/09654313.2012.651803.

Domènech, L., and D. Saurí. 2010. Socio-technical transitions in water scarcity contexts: Public acceptance of greywater reuse technologies in the Metropolitan Area of Barcelona. *Resources, Conservation and Recycling* 55: 53–62. doi:10.1016/j.resconrec.2010.07.001.

Eames, M., and J. Egmose. 2011. Community foresight for urban sustainability: Insights from the Citizens Science for Sustainability (SuScit) project. *Technological Forecasting and Social Change* 78(5): 769–784.

Elzen, B., F.W. Geels, and K. Green. 2004. *System innovation and the transition to sustainability: Theory, evidence and policy*. Cheltenham: Edward Elgar Publishing.

Ernstson, H., S.E. van der Leeuw, C.L. Redman, D.J. Meffert, G. Davis, C. Alfsen, and T. Elmqvist. 2010. Urban transitions: On urban resilience and human-dominated ecosystems. *AMBIO* 39: 531–545. doi:10.1007/s13280-010-0081-9.

Essletzbichler, J. 2012. Renewable energy technology and path creation: A multi-scalar approach to energy transition in the UK. *European Planning Studies* 20: 791–816. doi:10.1080/09654313.2012.667926.

Fainstein, S.S., and S. Campbell (eds.). 2012. *Readings in planning theory*, 3rd ed. Malden: Wiley-Blackwell.

Folke, C., S.R. Carpenter, B. Walker, M. Scheffer, T. Chapin, and J. Rockström. 2010. Resilience thinking: Integrating resilience, adaptability and transformability. *Ecology and Society* 15: 1–9.

Frantzeskaki, N., J. Wittmayer, and D. Loorbach. 2013. The role of partnerships in "realising" urban sustainability in Rotterdam's City Ports Area, the Netherlands. *Journal of Cleaner Production* 65(10): 406–417. doi:10.1016/j.jclepro.2013.09.023.

Geels, F.W. 2002. Technological transitions as evolutionary reconfiguration processes: A multi-level perspective and a case-study. *Research Policy* 31: 1257–1274. doi:10.1016/S0048-7333(02)00062-8.

Geels, F.W. 2010. Ontologies, socio-technical transitions (to sustainability), and the multi-level perspective. *Research Policy* 39: 495–510. doi:10.1016/j.respol.2010.01.022.

Geels, F.W. 2011. The role of cities in technological transitions – Analytical clarifications and historical examples. In *Cities and low carbon transitions*, ed. H. Bulkeley, V. Castán Broto, M. Hodson, and S. Marvin, 13–28. New York: Routledge.

Gunderson, L.H., and C.S. Holling (eds.). 2002. *Panarchy: Understanding transformations in human and natural systems*. Washington, DC: Island Press.

Guy, S., S. Marvin, and T. Moss (eds.). 2001. *Urban infrastructure in transition: Networks, buildings, plans*. London/Sterling: Earthscan Publication.

Guy, S., S. Marvin, W. Medd, and T. Moss (eds.). 2011. *Shaping urban infrastructures: Intermediaries and the governance of socio-technical networks*. London/Washington, DC: Earthscan.

Hagelskjær Lauridsen, E., and J. Stissing Jensen. 2013. The strictest energy requirements in the world: An analysis of the path dependencies of a self-proclaimed success. *Energy Policy* 53: 97–104. doi:10.1016/j.enpol.2012.10.014.

Hamann, R., and K. April. 2012. On the role and capabilities of collaborative intermediary organisations in urban sustainability transitions. *Journal of Cleaner Production* 50: 12–21. doi:10.1016/j.jclepro.2012.11.017.

Harding, A., and T. Blokland. 2014. *Urban theory*. Thousand Oaks: SAGE.

Healey, P. 1992. Planning through debate: The communicative turn in planning theory. *Town Planning Review* 62: 143–162.

Healey, P. 2007. *Urban complexity and spatial strategies: Towards a relational planning for our times*. London/New York: Routledge.

Hekkert, M.P., and S.O. Negro. 2009. Functions of innovation systems as a framework to understand sustainable technological change: Empirical evidence for earlier claims. *Technological Forecasting and Social Change* 76: 584–594. doi:10.1016/j.techfore.2008.04.013.

Hodson, M., and S. Marvin. 2010. Can cities shape socio-technical transitions and how would we know if they were? *Research Policy* 39: 477–485. doi:10.1016/j.respol.2010.01.020.

Hodson, M., S. Marvin, and H. Bulkeley. 2013. The intermediary organisation of low carbon cities: A comparative analysis of transitions in Greater London and Greater Manchester. *Urban Studies* 50: 1403–1422. doi:10.1177/0042098013480967.

Holling, C.S. 1973. Resilience and stability of ecological systems. *Annual Review of Ecology and Systematics* 4: 1–23.

Innes, J.E., and D.E. Booher. 1999. Consensus building and complex adaptive systems: A framework for evaluating collaborative planning. *Journal of the American Planning Association* 65: 412–423. doi:10.1080/01944369908976071.

Jabareen, Y. 2013. Planning the resilient city: Concepts and strategies for coping with climate change and environmental risk. *Cities* 31: 220–229. doi:10.1016/j.cities.2012.05.004.

Kemp, R., J. Schot, and R. Hoogma. 1998. Regime shifts to sustainability through processes of niche formation: The approach of strategic niche management. *Technology Analysis & Strategic Management* 10: 175–198. doi:10.1080/09537329808524310.

Khan J. 2013. What role for network governance in urban low carbon transitions? Journal of Cleaner Production 50:133–139. doi: 10.1016/j.jclepro.2012.11.045.

Lahlou, S. 2011. *System innovation for sustainability 4: Case studies in sustainable consumption and production – Energy use and the built environment*. Sheffield: Greenleaf Pub.

Loorbach, D. 2014. *To transition! Governance Panarchy in the new transformation*. Rotterdam: Inaugural address.

Loorbach, D., and J. Rotmans. 2010. The practice of transition management: Examples and lessons from four distinct cases. *Futures* 42: 237–246. doi:10.1016/j.futures.2009.11.009.

Lu, P., and D. Stead. 2013. Understanding the notion of resilience in spatial planning: A case study of Rotterdam, The Netherlands. *Cities* 35: 200–212. doi:10.1016/j.cities.2013.06.001.

Maassen, A. 2012. Heterogeneity of lock-in and the role of strategic technological interventions in urban infrastructural transformations. *European Planning Studies* 20: 441–460. doi:10.1080/09654313.2012.651807.

Mader, C. 2013. Sustainability process assessment on transformative potentials: The Graz model for Integrative development. *Journal of Cleaner Production* 49: 54–63. doi:10.1016/j.jclepro.2012.08.028.

Markard, J., R. Raven, and B. Truffer. 2012. Sustainability transitions: An emerging field of research and its prospects. *Research Policy* 41: 955–967. doi:10.1016/j.respol.2012.02.013.

Marsden, T. 2012. Sustainable place-making for sustainability science: The contested case of agri-food and urban–rural relations. *Sustainability Science* 8: 213–226. doi:10.1007/s11625-012-0186-0.

Mason, K., and M. Whitehead. 2012. Transition urbanism and the contested politics of ethical place making. *Antipode* 44: 493–516. doi:10.1111/j.1467-8330.2010.00868.x.

McCauley, S.M., and J.C. Stephens. 2012. Green energy clusters and socio-technical transitions: Analysis of a sustainable energy cluster for regional economic development in Central Massachusetts, USA. *Sustainability Science* 7: 213–225. doi:10.1007/s11625-012-0164-6.

McCormick, K., S. Anderberg, L. Coenen, and L. Neij. 2013. Advancing sustainable urban transformation. *Journal of Cleaner Production* 50: 1–11. doi:10.1016/j.jclepro.2013.01.003.

McGinnis, M.D., and E. Ostrom. 2014. Social-ecological system framework: Initial changes and continuing challenges. *Ecology and Society* 19(2): 30. doi:10.5751/ES-06387-190230.

Mieg, H.A. 2013. Introduction. In *Institutional and social innovation for sustainable urban development*, ed. H.A. Mieg and K. Töpfer, 1–21. Abingdon/New York: Routledge.

Mieg, H.A., and K. Töpfer (eds.). 2013. *Institutional and social innovation for sustainable urban development*. Abingdon/New York: Routledge.

Monstadt, J. 2009. Conceptualizing the political ecology of urban infrastructures: Insights from technology and urban studies. *Environment and Planning A* 41: 1924–1942. doi:10.1068/a4145.

Mulugetta, Y., T. Jackson, and D. van der Horst. 2010. Carbon reduction at community scale. *Energy Policy* 38: 7541–7545. doi:10.1016/j.enpol.2010.05.050.

Murdoch, J. 2006. *Post-structuralist geography: A guide to relational space*. London/Thousand Oaks: SAGE.

Næss, P., and N. Vogel. 2012. Sustainable urban development and the multi-level transition perspective. *Environmental Innovation and Societal Transitions* 4: 36–50. doi:10.1016/j.eist.2012.07.001.

Neal, S. 2013. Transition culture: Politics, localities and ruralities. *Journal of Rural Studies* 32: 60–69. doi:10.1016/j.jrurstud.2013.04.001.

Nevens, F., N. Frantzeskaki, L. Gorissen, and D. Loorbach. 2012. Urban transition labs: Co-creating transformative action for sustainable cities. *Journal of Cleaner Production* 50: 111–122. doi:10.1016/j.jclepro.2012.12.001.

Newton, P.W. (ed.). 2008. *Transitions: Pathways towards sustainable urban development in Australia*. Collingwood: CSIRO Pub.

North, P. 2013. Knowledge exchange, "impact" and engagement: Exploring low-carbon urban transitions. *The Geographical Journal* 179: 211–220. doi:10.1111/j.1475-4959.2012.00488.x.

North, P., and N. Longhurst. 2013. Grassroots localisation? The scalar potential of and limits of the "transition" approach to climate change and resource constraint. *Urban Studies* 50: 1423–1438. doi:10.1177/0042098013480966.

Peters, M.D., S. Fudge, and T. Jackson. 2012. *Low carbon communities: Imaginative approaches to combating climate change locally*. Cheltenham: Edward Elgar.
Pickett, S.T.A., C.G. Boone, B.P. McGrath, M.L. Cadenasso, D.L. Childers, L.A. Ogden, M. McHale, and J.M. Grove. 2013. Ecological science and transformation to the sustainable city. *Cities* 32: S10–S20. doi:10.1016/j.cities.2013.02.008.
Pincetl, S. 2012. Nature, urban development and sustainability – What new elements are needed for a more comprehensive understanding? *Cities* 29: S32–S37. doi:10.1016/j.cities.2012.06.009.
Portney, K.E. 2009. Sustainability in American cities: A comprehensive look at what cities are doing and why. In *Toward sustainable communities, Transition and transformations in environmental policy*. Cambridge, MA: MIT Press.
Quitzau, M.-B., B. Hoffmann, and M. Elle. 2012. Local niche planning and its strategic implications for implementation of energy-efficient technology. *Technological Forecasting and Social Change* 79: 1049–1058. doi:10.1016/j.techfore.2011.11.009.
Quitzau, M.-B., J.S. Jensen, M. Elle, and B. Hoffmann. 2013. Sustainable urban regime adjustments. *Journal of Cleaner Production* 50: 140–147. doi:10.1016/j.jclepro.2012.11.042.
Rauws, W.S., and G. de Roo. 2011. Exploring transitions in the peri-urban area. *Planning Theory and Practice* 12: 269–284. doi:10.1080/14649357.2011.581025.
Raven, R., J. Schot, and F. Berkhout. 2012. Space and scale in socio-technical transitions. *Environmental Innovation and Societal Transitions* 4: 63–78. doi:10.1016/j.eist.2012.08.001.
Reusswig, F. 2010. Sustainability transitions through the lens of lifestyle dynamics. In *Sustainable production consumption systems*, ed. L. Lebel, S. Lorek, and R. Daniel, 39–59. Dordrecht: Springer.
Roggema, R., T. Vermeend, and A. Dobbelsteen. 2012. Incremental change, transition or transformation? Optimising change pathways for climate adaptation in spatial planning. *Sustainability* 4: 2525–2549. doi:10.3390/su4102525.
Rotmans, J. 2006. A complex systems approach for sustainable cities. In *Smart growth and climate change: Regional development, infrastructure and Adaption*, ed. M. Ruth, 155–180. Cheltenham: Edward Elgar.
Rotmans, J., R. Kemp, and M. van Asselt. 2001. More evolution than revolution. Transition management in public policy. *Foresight – The Journal of Futures Studies, Strategic Thinking and Policy* 3: 15–31.
Rydin, Y., C. Turcu, S. Guy, and P. Austin. 2013. Mapping the coevolution of urban energy systems: Pathways of change. *Environment and Planning A* 45: 634–649. doi:10.1068/a45199.
Sassen, S. 1991. *The global city: New York, London, Tokyo*, 2nd ed. Princeton: Princeton University Press.
Scerri, A., and M. Holden. 2013. Ecological modernization or sustainable development? Vancouver's "greenest city action plan": The city as "manager" of ecological restructuring. *Journal of Environmental Policy & Planning* 16(2): 1–19. doi:10.1080/1523908X.2013.836962.
Scholz, R.W. 2011. *Environmental literacy in science and society: From knowledge to decisions*. Cambridge/New York: Cambridge University Press.
Schreuer, A., M. Ornetzeder, and H. Rohracher. 2010. Negotiating the local embedding of socio-technical experiments: A case study in fuel cell technology. *Technology Analysis & Strategic Management* 22: 729–743. doi:10.1080/09537325.2010.496286.
Seyfang, G., and A. Haxeltine. 2012. Growing grassroots innovations: Exploring the role of community-based initiatives in governing sustainable energy transitions. *Environment and Planning C: Government and Policy* 30: 381–400. doi:10.1068/c10222.
Shove, E., and G. Walker. 2010. Governing transitions in the sustainability of everyday life. *Research Policy* 39: 471–476. doi:10.1016/j.respol.2010.01.019.
Smith, A., and A. Stirling. 2010. The politics of social-ecological resilience and sustainable sociotechnical transitions. ed. The Resilience Alliance. *Ecology and Society* 15: 11.

Smith, A., and G. Seyfang. 2013. Constructing grassroots innovations for sustainability. *Global Environmental Change* 23: 827–829. doi:10.1016/j.gloenvcha.2013.07.003.
Smith, A., J.-P. Voß, and J. Grin. 2010. Innovation studies and sustainability transitions: The allure of the multi-level perspective and its challenges. *Research Policy* 39: 435–448. doi:10.1016/j.respol.2010.01.023.
Späth, P. 2012. Understanding the social dynamics of energy regions—The importance of discourse analysis. *Sustainability* 4: 1256–1273. doi:10.3390/su4061256.
Späth, P., and H. Rohracher. 2010. "Energy regions": The transformative power of regional discourses on socio-technical futures. *Research Policy* 39: 449–458. doi:10.1016/j.respol.2010.01.017.
Späth, P., and H. Rohracher. 2012. Local demonstrations for global transitions—Dynamics across governance levels fostering socio-technical regime change towards sustainability. *European Planning Studies* 20: 461–479. doi:10.1080/09654313.2012.651800.
STRN. 2010. *A mission statement and research agenda for the sustainability transitions research network*. Sustainability Transitions Research Network.
Taylor, P.J. 2012. Transition towns and world cities: Towards green networks of cities. *Local Environment* 17: 495–508. doi:10.1080/13549839.2012.678310.
Truffer, B., and L. Coenen. 2012. Environmental innovation and sustainability transitions in regional studies. *Regional Studies* 46: 1–21. doi:10.1080/00343404.2012.646164.
Truffer, B., E. Störmer, M. Maurer, and A. Ruef. 2010. Local strategic planning processes and sustainability transitions in infrastructure sectors. *Environmental Policy and Governance* 20: 258–269. doi:10.1002/eet.550.
UN SDSN. 2013. *The urban opportunity: Enabling transformative and sustainable development*. Bangalore/New York: UN Sustainable Development Solutions Network.
Uyarra, E., and S. Gee. 2012. Transforming urban waste into sustainable material and energy usage: The case of greater Manchester (UK). *Journal of Cleaner Production* 50(2): 101–110. doi:10.1016/j.jclepro.2012.11.046.
Van der Voorn, T. 2012. Combining backcasting and adaptive management for climate adaptation in coastal regions: A methodology and a South African case study. *Futures* 44: 346–364.
Vojnovic, I. 2014. Urban sustainability: Research, politics, policy and practice. *Cities* 41: S30–S44. doi:10.1016/j.cities.2014.06.002.
Wiek, A., R. Scholz, D. Lang, M. Stauffacher, and A. Walter. 2003. *Grundlagen des Transition-Managements von Regionen. Systemanalyse, Variantenkonstruktion, Bewertung. Ein Handbuch*. Zürich: Natural and Social Science Interface. Swiss Federal Institute of Technology.
Wiek, A., C. Binder, and R.W. Scholz. 2006. Functions of scenarios in transition processes. *Futures* 38: 740–766.
Wilson, G.A. 2012. Community resilience, globalization, and transitional pathways of decision-making. *Geoforum* 43: 1218–1231. doi:10.1016/j.geoforum.2012.03.008.
Wolfram, M., and R. Vogel. 2012. Governance and design of urban infostructures: Analysing Key socio-technical systems for the vulnerability and resilience of cities. *Raumforschung und Raumordnung* 70: 323–336. doi:10.1007/s13147-012-0169-8.
Young, R. 2010. The greening of Chicago: Environmental leaders and organisational learning in the transition toward a sustainable metropolitan region. *Journal of Environmental Planning and Management* 53: 1051–1068. doi:10.1080/09640568.2010.508948.

Exploring the Geography of Educational Segregation in Seoul, Korea

Up Lim, Ye Seul Choi, Chanyong Kim, and Donghyun Kim

Abstract This chapter examines changes in the spatial patterns of human capital segregation across neighborhoods in the Seoul metropolitan area during 2000–2010 and investigates the following three questions: (1) to what extent are highly educated individuals segregated from less-educated individuals across neighborhoods, (2) to what extent do highly educated and less-educated individuals live in isolated neighborhoods with individuals of similar educational status, and (3) to what extent can spatial clusters of highly educated or less-educated individuals be isolated across neighborhoods? Four major findings were obtained. First, the number and proportion of people with at least a college education increased markedly over time. Second, according to results of the dissimilarity index and the generalized dissimilarity index, the degree of segregation is highest for the group with more than a college education vis-à-vis the group with less than a high school education. Additionally, it is lowest for the group of high school graduates vis-à-vis the group with less than high school education in the Seoul metropolitan area over time. Third, the information theory index shows that the degree of diversity and human capital segregation steadily increased over time. Fourth, highly educated individuals tend to be clustered in the southern parts of the Seoul metropolitan area. By contrast, less-educated individuals were more likely to be concentrated in the mid-northern parts of the metropolitan area. The results of the empirical analysis in our study have implications for regional policies and can inform future research on the social processes and mechanisms of polarized educational segregation.

Keywords Educational segregation • Human capital • Neighborhood • Residential segregation • Seoul

---

U. Lim (✉) • Y.S. Choi • C. Kim
Department of Urban Planning and Engineering, Yonsei University, Seoul, Korea
e-mail: uplim@yonsei.ac.kr; yeseul.choi@yonsei.ac.kr; chanyongkim@yonsei.ac.kr

D. Kim
Korea Environment Institute, Sejong, South Korea
e-mail: donghyunkim@kei.re.kr

E. Kim, B.H.S. Kim (eds.), *Quantitative Regional Economic and Environmental Analysis for Sustainability in Korea*, New Frontiers in Regional Science: Asian Perspectives 25, DOI 10.1007/978-981-10-0300-4_2

1 Introduction

Korea has experienced rapid economic growth and development in the past decades. Historically, it has long been recognized that the higher level of human capital is the main engine of Korea's economic growth, because well-educated human capital resources are central for the absorption and application of advanced technologies from the developed countries. Moreover, from the beginning of the 1990s, increasingly keen competition for knowledge creation, changing job conditions in an environment of uncertain technological changes, and a rapid transition to a knowledge-intensive society have widely extended the concern for educational participation and investment in Korea. This tendency is emphasized by the education data from the Organization for Economic Cooperation and Development (OECD). According to the data recently announced by the OECD (2014), the college enrollment rate of high school graduates in Korea is 69 % in 2012, which is significantly higher than the OECD average of 58 %. Particularly, among the 34 member nations of the OECD, Korea's tertiary graduation rate of youth population between the ages of 25 and 34 years maintained its top position during the period 2007–2012 (OECD 2014).

While every corner of Korea benefited from an overall rapid growth of educational attainment, Seoul, the capital of Korea, has attracted far more college graduates than other metropolitan areas. The city of Seoul provides a powerful competitive advantage in attracting qualified human capital. This phenomenon is based on abundant job opportunities and outstanding residential and educational environment, infrastructure, and amenities. According to the 2010 Census of Population, 45.3 % of the population aged 25 years and over in Seoul is college educated, exceeding the national average of 35.3 %. Furthermore, in keeping with the increasing awareness about the benefits of educational attainment, the percentage of college-educated persons in Seoul has sharply increased from 23.0 % in 1990 to 45.3 % in 2010. However, the geographical concentration of highly educated groups in Seoul has both positive and negative effects. Although they have significantly contributed to stimulate Seoul's economic growth, the concentration of highly educated groups in Seoul has also led to neighborhood residential segregation by educational attainment. This has generated patterns of spatial imbalance and created disadvantaged neighborhood, which can be described with the figurative term "urban mosaic."

How does neighborhood residential segregation arise and how is it maintained or reinforced? One of the important areas of research in the field of social science has been the exploration of the relevant determinants of residential segregation. Therefore, a substantial body of research has attempted to discover which factors drive residential segregation across neighborhoods and has concluded that it is influenced by contextual factors specific to metropolitan areas (e.g., Mayer and Jencks 1989; Crane 1991; Massey and Denton 1993). Existing studies in residential segregation have mainly focused on the issue of residential segregation in racial/ethnic groups because these groups generally form distinct spatial patterns of over- and underrepresentation across residential areas. In the same vein, a large number of studies argue that mounting racial/ethnic inequality inherently deepens residential segregation as people sort across neighborhoods based on their education and income (e.g., Bayer and McMillan 2012; Bayer et al. 2014; Grigoryeva and Ruef 2015).

However, this general intuition still misses an important aspect of the effect of educational inequality on neighborhood residential segregation and the spatial dimension of educational segregation. Indeed, while racial/ethnic characteristics exacerbate residential segregation, educational attainment can also have a significant effect on residential segregation. For example, people of the well-educated class are likely to live in desirable areas that are distant from the neighborhoods occupied by their lower-class counterparts. In such circumstances, increased between-group disparities based on educational attainment can plausibly lead to greater levels of residential segregation and to the formation of highly educated residential enclaves within the metropolitan area. In particular, it is recognized that there has been a positive correlation between educational attainment and income level, which has significant implications for social and economic inequality across neighborhoods. In the same vein, Orfield and Lee (2005) noted that it is important to examine the dynamic aspect and the spatial dimension of educational segregation to explore the determinants of residential segregation. In response, we argue that the geography of educational segregation should be given considerably more attention.

Our focus on the spatial dimension of educational segregation as a channel deepening neighborhood residential segregation is motivated by two effects, human capital externalities and neighborhood effects, which have been widely discussed in the field of urban and regional studies. Educational segregation influences economic performance through the effect of neighborhood human capital externalities (Lim 2008). For example, segregation by educational attainment and the level of human capital might impede the sustainable cycle of economic growth by preventing knowledge transfers or other types of human capital externalities generated by formal and informal interaction between distinct classes of people. In particular, when residential segregation is substantial and the highly educated group (or the high-income group) is concentrated in residential enclaves across the neighborhoods, it is impossible for the less-educated group (or the low-income group) to settle in those neighborhoods. In such circumstances, the rise of educational segregation has the potential to diminish the economic opportunity of the less-educated group, which is generated by their highly educated counterparts.

Another reason that educational segregation has significant consequences is because of the residential neighborhood effects. People are more likely to be more influenced by their neighborhoods than by people who live at a greater distance from their residence. Most notably, increasing educational segregation appears to generate new inequalities in social and economic opportunity, creating a radical change in the geographic basis of human society. Due to the influence of adult role models, residing in socially isolated and disadvantaged neighborhoods has a negative effect on child and adolescent development. By contrast, the children of the highly educated persons will socialize increasingly with other children of well-educated and successful parents. As Brooks-Gunn et al. (1993) argued, we need to view neighborhoods as a potent source of unequal opportunity since neighborhoods impart considerable advantages and disadvantages to the children growing up in them (e.g., Vartanian and Gleason 1985; Wilson 1987; Crane 1991; Duncan and Raudenbush 1999; Leventhal and Brooks-Gunn 2000).

Despite the richness of the theoretical arguments about the various aspects and the geographical patterns of residential segregation, they are seldom captured in the dominant genre of empirical research. In particular, the issues of the spatial dimension of educational segregation have received only slight attention compared to that of racial/ethnic segregation (Schlitte 2012). While most highly developed countries' labor markets are marked by rising levels of segregation by human capital, analyses on residential segregation by disparate education levels are scarce, and empirical evidence on the possible effects and the spatial dimension of educational segregation is lacking. Particularly, previous research on residential segregation overlooks the geography of educational segregation and residential enclaves of human capital that are fixed in space and reinforce patterns of spatial inequality that shape and reshape the residential distribution of social classes.

Given the aforementioned concerns over the importance of the geography of educational segregation, this study explores changes in the spatial patterns of human capital segregation across neighborhoods in Seoul during 2000–2010. Using 2000, 2005, and 2010 census data, this paper improves our current knowledge of the geography of educational segregation by addressing the following questions: (1) to what extent are the less educated segregated from the highly educated, (2) to what extent do the less educated live in isolated neighborhoods with persons of similar status, and (3) to what extent can spatial clusters of the less educated as well as the highly educated be isolated across neighborhoods? To recast these questions, we use the dissimilarity index, the generalized dissimilarity index, and the information theory index to measure the level of residential segregation and to describe the distribution of residential segregation across different educational groups in neighborhoods within Seoul. We also use Moran's *I* statistics and a LISA cluster map to examine whether the level of educational attainment in a neighborhood is spatially correlated with the level of educational attainment in neighboring areas within the metropolitan area from 2000 to 2010.

2 Neighborhood Residential Segregation of Educational Attainment

The framework presented in this paper is theoretically based on the notion that educational inequality creates a divergence in the willingness to pay for neighborhood attributes. Distinct education groups have different preferences of the physical and social environments of neighborhoods, thereby inducing the phenomenon of neighborhood residential segregation by educational attainment. If so, why does neighborhood residential segregation of educational attainment matter? Many scholars in urban and regional studies have hypothesized that the issues of educational segregation and regional growth and development are closely connected. For instance, the increasing level of educational segregation has an impact on regional economic growth by spurring unemployment of the less educated. This phenomenon is closely related to the overall decline in the productivity levels of the educated group. This is because residential segregation by educational attainment reduces human capital externalities and neighborhood effects that bring positive social returns on the group with relatively insufficient schooling.

Localized human capital externalities, generated by the aggregate accumulation of human capital, are widely recognized as a key source of regional productivity. This argument is rooted in the seminal work of Lucas (1988), who provided a fundamental account of the role of knowledge spillovers. Lucas (1988) highlighted that social interaction between various groups gives rise to knowledge spillovers, and it persists with disparities in the economic development across regions. In line with this argument, Benabou (1993) and Durlauf (2004) also suggest that the aggregate accumulation of human capital can induce long-term growth of metropolitan areas because of the spillover effects of human capital externalities on economic growth. In order to investigate the complex mechanisms of economic growth, these studies explore human capital accumulation and its externalities, which contribute to the proliferation of new knowledge and technology. More recently, Schlitte (2012) suggested that localized human capital externalities that stem from the geographical accumulation of well-educated groups have a significant positive effect on the metropolitan economic productivity and outcomes. According to these works, economic development depends on advances in technology and scientific knowledge, and therefore, development depends on the accumulation of qualified human capital.

It is also possible that the spatial dimension of educational segregation influences individual economic performance through the effect of localized human capital externalities because educational attainment sorting strongly affects the geographical distribution of economic activities and social interactions. Given that economic networks and social interactions tend to be more intense between individuals located near one another than between individuals separated by long distances, people tend to be influenced by the composition of their neighborhoods (e.g., Quercia and Galster 2000; Sampson et al. 2002; Moretti 2004). Several studies have noted that the knowledge transfer occurring between highly and less-educated persons is a key source of human capital externalities and the likelihood of knowledge transfer rapidly decreases with distance (e.g., Audretsch and Feldman 2003; Ramos et al. 2010). In the same vein, Jovanovic and Rob (1989) and Glaeser (1999) theoretically argued that the geographical segregation between the highly and less-educated and skilled human capital decreases the chances for the latter to learn from the former. Additionally, Acemoglu (1996) showed that the presence of a large proportion of qualified local workers in a neighborhood encourages firms to increase its investment in the production of knowledge, which may increase the productivity of other workers. These arguments share the common consensus that the average level of human capital is a local public good (Lucas 1988; Schultz 1988; Azariadis and Drazen 1990).

Despite similar institutions and the same macroeconomic environment, the degree and the consequences of human capital externalities vary substantially across neighborhoods within the highly developed metropolitan areas. This is because human capital externalities and the consequent economic outcomes are strongly dependent on the distinct contexts of educational structure in the labor supply and the level of human capital segregation. Given the existence of localized human capital externalities, workers who live in labor markets with higher average levels of human capital tend to earn higher wages than workers in less-educated labor markets do, even after controlling for the characteristics of individual workers and dwellings (Rauch 1993). Consequently, we suggest that the productivity level of less-educated workers is positively influenced by human capital externalities arising from local human capital derived from highly educated and skilled workers. Hence, the spatial dimension of human capital segregation, including educational segregation, might affect economic performance of each neighborhood. Although the bulk of literature theoretically argues for the existence of a link between educational segregation and economic performance across the neighborhoods, only a few works have empirically explored the answer to the underlying question: how are people, who have different levels of educational attainment, spatially distributed across the neighborhoods in the metropolitan area? In the course of our discussion, we recast this question by using several segregation indexes. Additionally, a part of our discussion involves understanding the perspective that each index brings to the discussion on educational segregation.

Another explanation of the link between educational segregation and individual economic performance is the neighborhood effect on individual social and economic outcomes. However, there is currently no precise and uniform definition of the concept of the neighborhood and neighborhood effects. Referring to the definition by Galster (2001), a neighborhood can be defined as "the bundle of spatially based attributes associated with clusters of residence, sometimes in conjunction with other land uses." As a multidimensional bundle consisting of everything from structures and topography to demography, public services, and social interactions, the unifying feature of these attributes is that they are spatially based and measured only after a particular location has been specified (Galster 2001). The neighborhood environment is crucial to the well-being of households and is of comparable importance to the dwelling itself in shaping expressions of residential satisfaction. As Galster and Killen (1995) argued, the spatially variant geography of metropolitan opportunity has an important effect on youth's decisions regarding education, fertility, work, and crime.

Historically, urban scholars have tended to establish whether geography and the spatial dimension of individuals are related to the individuals' outcomes and, if so, what the nature of this relationship is. Over the years, various attempts have been made to recast this question, and it is noted that the effect of metropolitan-wide residential segregation on individual outcomes is one effective description. In the same vein, residential segregation has been the topic of several past studies (e.g., Mayer and Jencks 1989; Crane 1991; Massey and Denton 1993 among others). In broad terms as an outcome, residential segregation simply implies the uneven distribution of different population groups within a local and regional housing system (Gordon and Monastiriotis 2006). The classic literature by Massey and Denton (1993) demonstrated the mechanisms of the formation of concentrated poverty in minority neighborhoods. They argued that residential segregation, most notably racial segregation, reinforces the effects of social and economic deprivation in each community. In this context, residential segregation, as a metropolitan-wide phenomenon, is thought to be the instrument through which neighborhood disparities are created and maintained.

The central theme in the residential segregation literature is that there are certain social processes or mechanisms that take place at the neighborhood level that affect outcomes for individuals. The theme is primarily rooted in William Julius Wilson's seminal book *The Truly Disadvantaged* (1987). The book fueled a lively scholarly controversy on how living in disadvantaged neighborhood has negative effects on human development. A number of social scientists began to explicitly theorize and directly measure the bearing of neighborhood social processes or mechanisms on the well-being of individuals, particularly children and adolescents (Sampson et al. 2002). Concern over residential segregation has motivated empirical research geared toward distilling and analyzing the determinants of the dynamics of neighborhood inequality (Massey 1996; Timberlake 2002; Sampson 2009). Expanding Massey's (1996) argument on the age of extremes, Swanstrom et al. (2002) maintain that growing class segregation in metropolitan areas has enormous negative social and political consequences.

The main consensus of the previous empirical literature on neighborhood residential segregation is that racial/ethnic inequality has played a key role in deepening residential segregation. Despite a strong influence of educational segregation on the geographical segregation and spatial inequality across the neighborhoods, few scholars have investigated the geography and the spatial dimension of educational segregation. In this way, there are only very few empirical investigations that explore how residential enclaves of human capital can be identified across neighborhoods in a metropolitan area. In this paper, our argument is based on the work of Sampson (2009), who stated, "disadvantage is not encompassed in a single characteristic but rather is a synergistic composite of social factors that mark the qualitative aspects of growing up in truly disadvantaged neighborhoods." This matter is especially relevant because the geography of educational segregation tends to be fixed in space and reinforces patterns of spatial inequality that shape and reshape the residential distribution of social classes. This study will provide new insights into social processes or mechanisms underlying the residential landscapes of segregation.

3 Data and Methods

This study uses 424 dong areas in 25 districts of the Seoul metropolitan area, as shown in Fig. 1. The data for the educational attainment for the population aged 25 years and above was gathered from the 2000, 2005, and 2010 Census of Population files. To capture the patterns of educational segregation in Seoul, this study divides educational attainment into four mutually exclusive educational groups: less than high school (<HS), high school graduate (HS), college graduate (CG), and more than college (>CG). This study considers college-educated individuals as the population of high-human capital individuals because the recent empirical literature on human capital externalities has focused on these highly educated groups as key actors in creating spillover effects (Rosenthal and Strange 2008). Therefore, the two college-educated groups (i.e., college graduate and more than college) are considered as the group of highly educated individuals in our study.

Fig. 1 424 dongs in 25 districts of the Seoul metropolitan area in 2010. (1) Jongno, (2) Jung, (3) Yongsan, (4) Seongdong, (5) Gwangjin, (6) Dongdaemun, (7) Jungnang, (8) Seongbuk, (9) Gangbuk, (10) Dobong, (11) Nowon, (12) Eunpyeong, (13) Seodaemun, (14) Mapo, (15) Yangcheon, (16) Gangseo, (17) Guro, (18) Geumcheon, (19) Yeongdeungpo, (20) Dongjak, (21) Gwanak, (22) Seocho, (23) Gangnam, (24) Songpa, and (25) Gangdong

The evaluation of the level of residential segregation of different educational groups has long been a major issue for social researchers involved in urban and regional policies and, to a greater extent, urban studies. Residential segregation has been extensively studied with various measures in the existing literature (Duncan and Duncan 1955). Most measures of residential segregation, including aspatial (Massey and Denton 1988) and spatial (Reardon and O'Sullivan 2004; Reardon et al. 2008) indexes, are based on the premise that equates proximity with potential interaction between individuals from different groups. As one of the aspatial segregation indexes, the index of which is a measure of evenness and is the most widely used measurement of residential segregation. The dissimilarity index measures the fraction of the two educational groups that would need to change residence for each neighborhood to have the same fraction of the group as the overall metropolitan area (Duncan and Duncan 1955; Cutler et al. 1999). This index permits the comparison of two educational groups' spatial distributions across the spatial units of a metropolitan area (Apparicio et al. 2013). The dissimilarity index is defined as

$$D = \frac{1}{2}\sum_{i=1}^{n}\left|\frac{x_i}{X} - \frac{y_i}{Y}\right| \quad (1)$$

where x_i and y_i are the numbers of individuals in a specific educational group in neighborhood i and X and Y are the total number of individuals in a specific educational group in the metropolitan area as a whole. According to Massey and Denton (1988), evenness is maximized and segregation is minimized when all geographic units have the same relative number of minority and majority members as the city as a whole and vice versa. This index varies between 0 (no segregation) and 1 (perfect segregation).

However, D has severe limitations in reflecting the spatial dimension of residential segregation. A major problem with the formulation of D is that it treats enumeration unit boundaries as actual boundaries separating people across units and thus considers these boundaries as inhibiting interaction or mixing among groups (Wong 2005). Thus, by adjusting the index of dissimilarity, we try to capture the interaction information indirectly by using the concept of composite population counts, which treat different population groups in neighboring areas as if they are in the same area (Wong 2005). The composite population count of the specific educational group in geographic unit i is defined as

$$c\_x_i = \sum_{j=1}^{n} d(x_j) \quad (2)$$

where x_j is the specific educational group population count in geographic unit j, $d(\cdot)$ is a function defining the neighborhood of i, and j can be the same as i. c_x_i is the composite-specific educational group population count of x_i in geographic unit i. Similarly, we can define c_y_i, the composite population count of another educational group in geographic unit i, in the same manner as in Eq. (2). The traditional approach to specifying the function of neighborhood relies on the spatial arrangement of the geographic units, designing geographic units as neighbors when they share a common border. For example, $d(x_j)$ has a value of 1 if geographic units i and j share a common border, and 0 otherwise. These composite counts involve populations in the neighboring unit i such that people in different educational groups within the neighborhood of i can interact as if they are in the same geographic unit i. Based on these composite population counts, the index of generalized spatial segregation *GD* as a spatial segregation index, proposed by Wong (2005), can be defined as

$$GD = \frac{1}{2}\sum_{i=1}^{n}\left|\frac{c\_x_i}{\sum_{i=1}^{n} c\_x_i} - \frac{c\_y_i}{\sum_{i=1}^{n} c\_y_i}\right| \tag{3}$$

where the denominators are the total composite population counts of two different educational groups for the metropolitan area. The *GD* index is bounded between 0 and 1, indicating no spatial segregation and perfect spatial segregation, respectively. By using this index, we can account for the potential spatial interaction between educational groups among neighboring units.

Although the dissimilarity indexes (*D*) are most widely used in the segregation literature, it is limited to the traditional dual-group context because it calculates segregation for only two groups at a time (e.g., Duncan and Duncan 1955; James and Taeuber 1985; Massey and Denton 1988; Massey and Denton 1993). In addition to looking at the segregation of each educational group separately, this study also examines educational segregation in the multigroup context by using the information theory index because the information theory index can measure the spatial distribution of multiple groups simultaneously. Compared to the dissimilarity index, the information theory index can also be decomposed into its components (Hårsman and Quigley 1995; Reardon and Firebaugh 2002). For these reasons, we use the information theory index for this study. Theil's information theory index is the weighted average deviation of each neighborhood's entropy from that of the metropolitan-wide diversity, expressed as a fraction of the metropolitan area's total entropy:

$$H = \sum_{i=1}^{n}\frac{t_i(E - E_i)}{ET} \tag{4}$$

where t_i represents the total population aged 25 years and above in neighborhood i and T is the metropolitan area population aged 25 years and above. E and E_i refer to metropolitan area entropy and neighborhood i's entropy, respectively. Metropolitan area entropy is calculated as $E = \sum \pi_r \ln(1/\pi_r)$, where π_r refers to the proportion of a particular educational group in the whole metropolitan area population. A neighborhood within the metropolitan area would analogously have its entropy score defined as $E_i = \sum \pi_{ri} \ln(1/\pi_{ri})$, where π_{ri} refers to the proportion of a particular educational group in the population of neighborhood i. Using the information theory index, we measure how evenly educational groups are distributed across neighborhoods within the metropolitan area and can decompose it into its components. The information theory index ranges from 0 (when each neighborhood has the same educational diversity as the entire metropolitan area) to 1 (when each neighborhood contains members of only a single educational group).

Although the dissimilarity index and the information theory index show the extent to which educational groups are evenly distributed across neighborhoods in the metropolitan area, they provide no spatial information on the geography of educational segregation in the metropolitan area, masking considerable underlying spatial complexity. There is a rising need to explore the spatial correlation between educational groups and the patterns of residential clustering in the metropolitan area over time to broaden our understanding of the spatial structures of educational segregation. In this context, the current study applies the methods of exploratory spatial data analysis (ESDA) to investigate whether the concentration of a specific educational group in a neighborhood is spatially correlated with the concentration of the same educational group in the neighboring areas and to further examine whether spatial clusters of a specific educational group can be isolated across neighborhoods in the metropolitan area. This study first considers a global measure of spatial autocorrelation, which is typically based on Moran's I statistic (Cliff and Ord 1981; Upton and Fingleton 1985). Moran's I statistic is the most commonly used method of spatial autocorrelation and is defined as follows:

$$I = \left(\frac{n}{s_0}\right) \frac{\sum_{i=1}^{n} \sum_{j=1}^{n} w_{ij} z_i z_j}{\sum_{i=1}^{n} {z_i}^2} \tag{5}$$

where z_i and z_j are the deviation of an attribute for the geographic unit i and unit j from its mean ($x_i - \overline{X}$), respectively, w_{ij} is the spatial weight between the geographic units i and j, and n is equal to the total number of geographic units. The element of spatial weight w_{ij} is 1 if the geographic units i and j have a common border, and 0 otherwise, and s_0 is a normalizing factor that is equal to the sum of the elements of the weight matrix (i.e., $s_0 = \sum \sum w_{ij}$) (Anselin 1995). Moran's I statistic varies from -1 to 1, indicating negative spatial autocorrelation and positive spatial autocorrelation, respectively. If Moran's I statistic has a positive value, then similar values are more likely to cluster themselves, and vice versa.

Moran's I statistic provides insight into the degree of spatial clustering. However, it does not fully identify the local structure of spatial autocorrelation because it is a measure of a global statistic that is calculated as the sum of values for an entire study area. Thus, we focus on a local indicator of spatial autocorrelation (LISA), the local version of Moran's I statistic, to identify local spatial clusters (i.e., hot spots of human capital) in the metropolitan area. The local Moran's I_i statistic for each neighborhood i can be defined by the following formula:

$$I_i = z_i \sum_{j=1}^{n} w_{ij} z_j \tag{6}$$

where z_i and z_j are the deviation of an attribute for the geographic units i and j from its mean $(x_i - \overline{X})$, respectively. w_{ij} is the spatial weight between the geographic units i and j. The element of spatial weight w_{ij} is 1 if the geographic units i and j have a common border, and 0 otherwise. n is equal to the total number of geographic units, and the summation over j is such that only neighboring values of j are included. A positive I_i value, which is known as spatial clustering of similar values, indicates that similar values are spatially clustered, whereas a negative I_i value indicates that the geographic unit is surrounded by its neighboring areas with dissimilar values, that is, there is spatial clustering of dissimilar values (Anselin 1995).

The local Moran's I statistic can be visualized in the form of a LISA cluster map. A LISA cluster map is classified into four types of local spatial association between the geographical area and its neighbors: (1) HH, high-high association (high values surrounded by high values); (2) LL, low-low association (low values surrounded by low values); (3) HL, high-low association (high values surrounded by low values); and (4) LH, low-high association (low values surrounded by high values). The HH and LL associations refer to a positive spatial autocorrelation, and the HL and LH associations indicate a negative spatial autocorrelation. Our study uses a LISA cluster map to show the changes in spatial clustering patterns in Seoul during 2000–2010.

4 Empirical Findings

Table 1 provides some basic statistics for educational attainment in the Seoul metropolitan area over the 2000–2010 period. As shown in Table 1, the total population aged 25 years and above increased from 2000 to 2010. During the study period, while the number and proportion of individuals with less than a college education steadily decreased, the number of college-educated individuals and the proportion of college-educated individuals in the population aged 25 years and above increased significantly. Table 2 presents both the dissimilarity indexes (*D*) and the generalized dissimilarity index (*GD*) for all pairs of educational groups in the metropolitan area from 2000 to 2010. The dissimilarity indexes for all pairs of educational groups measure the degree of evenness of distribution for the all pairs of educational groups in the metropolitan area over the 2000–2010 period.

The dissimilarity indexes for all pairs of educational groups as shown in Table 2 reveal that segregation is highest for the group of more than college vis-à-vis that of

Table 1 Basic educational attainment statistics in Seoul, 2000–2010

	2000		2005		2010	
Less than high school	1,638,880	(26.1)	1,421,961	(21.4)	1,346,647	(19.4)
High school graduate	2,576,291	(40.9)	2,511,960	(37.8)	2,416,483	(34.7)
College graduate	1,862,149	(29.6)	2,414,289	(36.3)	2,744,879	(39.5)
More than college	214,051	(3.4)	305,922	(4.5)	448,907	(6.4)
Total	6,291,371	(100.0)	6,654,132	(100.0)	6,956,916	(100.0)

Note: Percentages are in parentheses

Table 2 Summary of dissimilarity indexes and generalized dissimilarity indexes, 2000–2010

(a) Dissimilarity indexes (*D*)									
	2000			2005			2010		
	HS	CG	>CG	HS	CG	>CG	HS	CG	>CG
<HS	0.104	0.278	0.433	0.094	0.260	0.421	0.094	0.258	0.417
HS		0.201	0.371		0.190	0.367		0.185	0.360
CG			0.189			0.200			0.199
(b) Generalized dissimilarity index (*GD*)									
	2000			2005			2010		
	HS	CG	>CG	HS	CG	>CG	HS	CG	>CG
<HS	0.079	0.217	0.341	0.067	0.204	0.342	0.060	0.175	0.299
HS		0.157	0.291		0.154	0.298		0.125	0.258
CG			0.147			0.157			0.144

Note: <HS less than high school, HS high school graduate, CG college graduate, >CG more than college

less than high school, followed by the group of more than college vis-à-vis that of high school graduates and the group of college graduates vis-à-vis that of less than high school. Compared to this result, the lowest dissimilarity index score, which indicates a relatively even distribution, is for the group of less than high school vis-à-vis that of high school graduates. As the dissimilarity index increases, the degree of segregation increases. Thus, the findings reveal that the degree of segregation is highest for the group of more than college vis-à-vis that of less than high school (0.433 in 2000, 0.421 in 2005, and 0.417 in 2010) and is lowest for the group of high school graduates vis-à-vis that of less than high school (0.104 in 2000, 0.094 in 2005, and 0.094 in 2010) in the metropolitan area for the entire period.

However, people in neighboring geographic units can interact and are not perfectly segregated by administrative boundaries if there are no effective physical or administrative barriers stopping people from interacting. To capture the potential spatial interaction between educational groups among geographic units explicitly and the spatial dimension of segregation effectively, we use another index, the *GD*, which adjusts the level of *D* by using the concept of composite population counts (Wong 2005). As shown in Table 2, the values of *GD* are lower than the dissimilarity index, which refers to the aspatial *D*. This is because the potential interaction between the two educational groups is accounted for by *GD*. Moreover, similar to the results for *D*, the values of *GD* also show that the degree of segregation is highest for the group of more than college vis-à-vis that of less than high school (0.341 in 2000, 0.342 in 2005, and 0.299 in 2010). The results also show that the degree of segregation is lowest for the group of high school graduates vis-à-vis that of less than high school (0.079 in 2000, 0.067 in 2005, and 0.060 in 2010) in the metropolitan area for the entire period.

Table 3 includes the results for the diversity and multigroup segregation measures. The diversity and multigroup segregation measures show that diversity increased over the period by 0.038 and the degree of segregation remained the same or changed slightly. When decomposing the multigroup segregation index into the component groups, we find that segregation is greatest for the group of more than college over the period. That is, when looking at the segregation of people of each educational group vis-à-vis the rest of the population, segregation is highest for the most educated group. In contrast, the group of high school graduates was less segregated than other educational groups. Moreover, the groups of less than high school, high school, and more than college are more unevenly distributed across neighborhoods in the metropolitan area in 2010 than in 2000. The results reveal that overall increases in segregation over the period are mainly due to increases in segregation of the group of more than college, the group of high school, and the group of less than high school, while the degree of segregation in the group of college educated decreased.

Table 4 presents the results of the segregation decomposition into the 25 component districts of Seoul. As shown in Table 4, the values of information theory index remained the same or changed slightly for nearly all of the districts in the metropolitan area. These results reveal that overall educational segregation in the metropolitan area remained the same or changed marginally over the 2000–2010 period. By decomposing the information theory index of Seoul into the 25 component districts, we find that educational segregation is greatest for Yangcheon over the period under consideration, followed by Jongno and Yongsan in 2000, Yeongdeungpo and Jongno in 2005, and Songpa and Yeongdeungpo in 2010. Among the top ten districts along with educational segregation in 2010, Songpa, Gangnam, and Guro have experienced upward movement, while Yangcheon, Jongno, Yongsan, Gangdong, and Seocho experienced downward movement over the period. The Spearman's rank-order correlation coefficient, $r_s = 0.933$, of

Table 3 Entropy and Theil's information theory indexes, 2000–2010

	2000	2005	2010
Entropy diversity score	1.191	1.207	1.229
Multigroup information theory index	0.021	0.021	0.022
<HS information theory index	0.023	0.022	0.027
HS information theory index	0.006	0.010	0.012
CG information theory index	0.026	0.020	0.016
>CG information theory index	0.045	0.045	0.047

Note: <HS less than high school, HS high school graduate, CG college graduate, >CG more than college

Table 4 Decomposition of Theil's information theory index, 2000–2010

	2000	2005	2010
Seoul	0.021	0.021	0.022
Yangcheon	0.062 (1)	0.060 (1)	0.057 (1)
Jongno	0.047 (2)	0.044 (3)	0.043 (4)
Yongsan	0.047 (3)	0.043 (4)	0.039 (6)
Songpa	0.044 (4)	0.042 (5)	0.046 (2)
Yeongdeungpo	0.044 (5)	0.045 (2)	0.044 (3)
Gangdong	0.035 (6)	0.028 (9)	0.020 (10)
Seocho	0.033 (7)	0.031 (8)	0.026 (8)
Gwangjin	0.032 (8)	0.033 (7)	0.031 (7)
Gangnam	0.030 (9)	0.039 (6)	0.041 (5)
Dobong	0.029 (10)	0.022 (10)	0.019 (11)
Nowon	0.023 (11)	0.019 (14)	0.017 (14)
Jung	0.021 (12)	0.021 (12)	0.015 (15)
Gwanak	0.021 (13)	0.021 (13)	0.018 (12)
Guro	0.019 (14)	0.022 (11)	0.022 (9)
Mapo	0.018 (15)	0.012 (18)	0.011 (18)
Seongbuk	0.018 (16)	0.016 (16)	0.018 (13)
Seongdong	0.016 (17)	0.019 (15)	0.015 (16)
Gangbuk	0.012 (18)	0.007 (24)	0.010 (19)
Gangseo	0.011 (19)	0.014 (17)	0.014 (17)
Seodaemun	0.011 (20)	0.011 (19)	0.010 (21)
Dongdaemun	0.009 (21)	0.008 (23)	0.007 (25)
Jungnang	0.009 (22)	0.010 (20)	0.008 (23)
Eunpyeong	0.008 (23)	0.005 (25)	0.009 (22)
Dongjak	0.007 (24)	0.008 (22)	0.009 (24)
Geumcheon	0.007 (25)	0.010 (21)	0.010 (20)

educational segregation between 2000 and 2010 suggests a very high stability over time in the hierarchy of educational segregation among all districts in the metropolitan area. On the other hand, it is still worth paying attention to several upward and downward movements among the districts. In line with the results, this analysis casts new light on the changing geographical patterns of segregation of educational attainment among the districts within the metropolitan area during 2000–2010.

To gain a deeper understanding of the changes in the spatial geography of educational segregation in the Seoul metropolitan area over time, we examine whether the clustering of highly educated and less-educated individuals in a geographic area was spatially correlated with that of the same educational group in neighboring areas. Table 5 provides information on Moran's I statistics and spatial clustering patterns of educational attainment by district over the 2000–2010 period. Moran's I statistic is calculated to explore the residential patterns of segregation and spatial clustering for each educational group in the Seoul metropolitan area during the 2000–2010 period. Moran's I statistic for all educational groups in the metropolitan area as a whole increased during 2000–2005, while

Table 5 Moran's *I* statistics, 2000–2010

	2000				2005				2010			
	<HS	HS	CG	>CG	<HS	HS	CG	>CG	<HS	HS	CG	>CG
Seoul	0.545*	0.471*	0.551*	0.471*	0.553*	0.545*	0.574*	0.495*	0.478*	0.465*	0.427*	0.457*
(1) Jongno	0.381*	0.311*	0.424*	0.355*	0.569*	0.545*	0.620*	0.508*	0.351*	0.176*	0.064	0.094
(2) Jung	0.317*	0.044	0.304*	0.272*	0.229*	0.059	0.212*	0.130	0.010	−0.349*	−0.023	0.023
(3) Yongsan	−0.136	−0.211*	−0.156	−0.186	0.019	−0.167	−0.031	−0.029	0.042	0.101	−0.034	−0.075
(4) Seongdong	0.009	−0.011	0.019	0.057	0.098	−0.021	0.046	0.135*	0.078	−0.071	0.042	0.084
(5) Gwangjin	0.275*	0.114*	0.293*	0.274*	0.296*	0.187*	0.337*	0.204*	0.191*	−0.090	−0.039	−0.022
(6) Dongdaemun	−0.020	−0.199	−0.043	0.018	0.129	0.167*	0.214*	0.096	−0.080	−0.045	−0.045	0.035
(7) Jungnang	0.142	0.125*	0.169*	0.143*	0.164*	0.120	0.147	0.139*	−0.082	−0.321*	−0.062	−0.008
(8) Seongbuk	0.319*	0.274*	0.341*	0.393*	0.469*	0.286*	0.461*	0.260*	0.073	0.092	0.356	0.339*
(9) Gangbuk	−0.086	−0.134	−0.087	−0.086	0.011	−0.004	0.056	−0.037	0.071	0.010	−0.137	−0.051
(10) Dobong	0.082	−0.001	0.061	0.176*	0.118	0.354*	0.232*	0.199*	0.043	−0.064	0.023	0.168*
(11) Nowon	0.008	−0.120	−0.013	−0.001	0.328*	0.161*	0.264*	0.210*	0.221*	0.077	−0.100	−0.061
(12) Eunpyeong	−0.096	0.020	−0.096	0.039	−0.093	−0.134	−0.128	−0.035	−0.255*	−0.077	0.108*	0.026
(13) Seodaemun	0.296*	0.280*	0.335*	0.340*	0.264*	0.277*	0.304*	0.249*	0.016	−0.234	0.091	0.113
(14) Mapo	0.164*	−0.094	0.131	0.077	−0.091	0.039	−0.028	−0.014	−0.048	−0.093	−0.060	0.002
(15) Yangcheon	0.324*	0.245*	0.366*	0.372*	0.221*	0.244*	0.262*	0.191*	0.017	−0.057	0.174*	0.238*
(16) Gangseo	−0.272*	−0.151	−0.252*	−0.244*	−0.126	−0.069	−0.113	−0.162	−0.059	0.117	−0.173	−0.260*
(17) Guro	−0.093	0.128*	−0.089	0.087	−0.340*	−0.192*	−0.348*	−0.180	−0.316*	−0.061	0.109	0.061
(18) Geumcheon	0.098	−0.062	0.041	0.093	0.094	0.089	0.052	0.194*	0.064	−0.485*	−0.262*	−0.085
(19) Yeongdeungpo	0.229*	0.266*	0.281*	0.320*	0.417*	0.238*	0.349*	0.328*	0.071	−0.129	0.130*	0.159*
(20) Dongjak	0.041	−0.000	0.044	0.066	−0.067	0.139	0.000	0.052	−0.062	0.009	0.206	0.115
(21) Gwanak	0.219*	0.058	0.243*	0.133*	0.023	0.137*	0.094	0.148*	−0.292*	−0.128	0.202*	0.081
(22) Seocho	0.233*	0.009	0.232*	0.229*	0.297*	0.267*	0.313*	0.222*	0.098	0.026	0.136	0.148*
(23) Gangnam	0.002	−0.018	−0.024	0.052	0.107	0.088	0.101	0.110	0.121*	0.050	−0.032	0.016
(24) Songpa	0.007	0.087	0.049	0.074	0.132	0.209*	0.168*	0.200*	0.150*	0.159*	0.175*	0.152*
(25) Gangdong	−0.097	0.061	−0.085	0.022	0.235*	0.137*	0.243*	0.198*	−0.012	−0.178	−0.130	−0.085

Notes: <HS less than high school, HS high school graduate, CG college graduate, >CG more than college
*Significant at the 0.10 level

Moran's I statistic for all educational groups in the metropolitan area decreased during 2005–2010. As seen in Table 5, the residential pattern of segregation for each educational group in the metropolitan area as a whole is spatially clustered and tends to be more clustered during 2000–2005. On the other hand, the residential pattern of segregation for each educational group tends to be less clustered during 2005–2010.

Another interesting finding is the trend in spatial clustering of human capital over the 2000–2010 period as suggested by the LISA cluster map. We compute the local Moran's I_i statistic for each neighborhood to determine the local spatial clusters of a specific educational group across neighborhoods and visualize it in the form of a LISA cluster map. A positive local Moran's I_i statistic value refers to spatial clustering of similar values, whereas a negative value indicates spatial clustering of dissimilar values between areas. As previously mentioned, significant HH quadrants or LL quadrants indicate a local spatial cluster. Figure 2 shows that the significant HH neighborhoods for the highly educated groups are located in southern areas, such as the Seocho, Gangnam, and Songpa districts, whereas the significant LL neighborhoods are located in the midwestern portion of the metropolitan area, such as the Jung, Dongdaemun, Jungnang, and Gangbuk districts. For the highly educated groups, Mokdong located in the Yangcheon district, which is one of the significant HH neighborhoods, emerged as a hot spot of human capital over time.

Compared to these results, the spatial clusters of significant HH for the less-educated groups are concentrated in the northern area, such as the Jungnang, Gangbuk, Dobong, and Nowon districts. These clusters are maintained or reinforced over time. In contrast, the significant LL neighborhoods for the less-educated groups are spatially clustered in the mid-southern area of Seoul, such as the Jongno, Yongsan, Seocho, Gangnam, and Songpa districts from 2000 to 2010. These results for the spatial patterns of educational segregation in the Seoul metropolitan area show a highly uneven geographical distribution of educational attainment. Similarly, the highly educated groups tend to be spatially clustered in certain parts of the southern areas, such as the Seocho and Gangnam districts. In contrast, the group of less-educated individuals has concentrated in the northern parts of the Seoul metropolitan area, such as the Dongdaemun, Jungnang, Gangbuk, Dobong, and Nowon districts during this period. As this spatial trend persisted over time, the geographical polarization of human capital was markedly fixed, as shown in Fig. 2.

Based upon these findings, the spatial segregation of human capital seems systematically linked to other dimensions of socioeconomic isolation and residential separation, and these patterns of spatial segregation may be associated with uneven urban development policies for the southern parts of Seoul metropolitan area, such as the Gangnam, Seocho, and Songpa districts (Lim 2008). This spatial segregation of human capital is closely associated with the behavioral pattern of the highly educated. The highly educated individuals tend to be attracted to the most desirable environments, where affluent social and educational resources tend to aggregate, especially in the Gangnam, Seocho, and Songpa districts, thereby

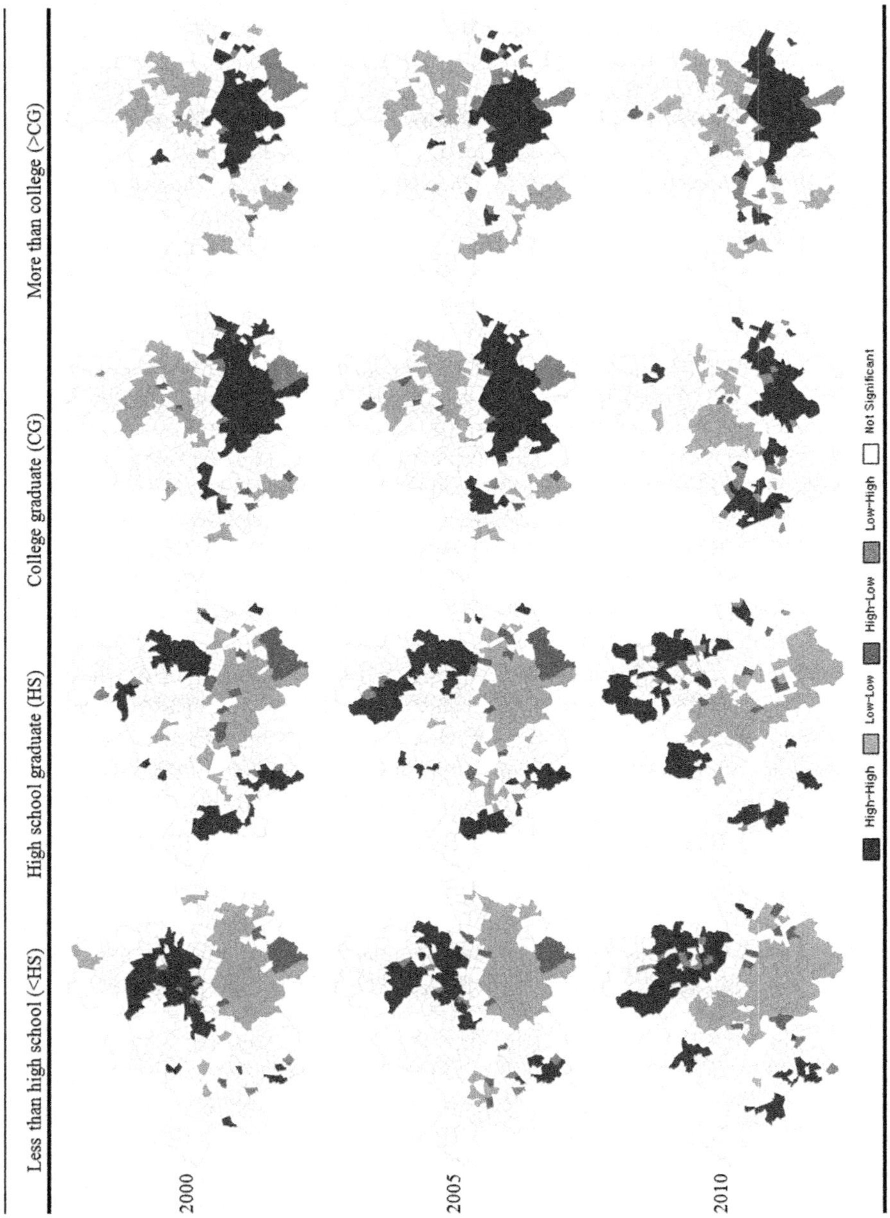

Fig. 2 LISA cluster map by educational group in Seoul, 2000–2010

reinforcing the spatial segregation of human capital. This uneven pattern of urban development and growth is regarded as a major cause of the social stratification and differentiation of groups in urban spaces and is thus a fundamental contributor to the segregated landscape (Smith 1991).

5 Conclusions

This study investigates changes in the spatial patterns of educational segregation across neighborhoods in the Seoul metropolitan area during 2000–2010 and examines whether each educational group in a neighborhood is spatially correlated with the same groups or isolated across neighborhoods within the metropolitan area. To gain a better understanding, we first analyze educational segregation from a neighborhood perspective over the 2000–2010 period. The major finding is that both the number of college-educated individuals and the proportion of college-educated individuals in the population aged 25 years and above increased significantly over time. According to the results of the aspatial dimension of segregation measured using the dissimilarity index, the generalized dissimilarity index, and the information theory index, the degree of segregation is highest for the group of more than college vis-à-vis that of less than high school, followed by the group of more than college vis-à-vis that of high school graduates. In contrast, the degree of segregation for the group of less than high school vis-à-vis that of high school graduates is lowest.

In addition, we conduct spatial analysis to further examine the spatial distribution of human capital in the metropolitan area using Moran's I statistic and a LISA cluster map. The results of the spatial analysis show a residential pattern with spatial clustering in certain areas. Another interesting finding is that most of the HH neighborhoods for the group of highly educated individuals are located in the southern parts of Seoul, such as the Seocho, Gangnam, Songpa, and Yangcheon districts during the period. By contrast, the HH neighborhoods for less-educated groups (i.e., less than high school and high school graduates) tend to be located in the northern parts of the Seoul metropolitan area, such as the Jungnang, Gangbuk, Dobong, and Nowon districts.

The results suggest that highly educated or less-educated individuals might have lived in educationally segregated enclaves and were therefore more exposed to similarly educated people, as shown by the geographic polarization of educational landscape in the results. The polarized geography of educational segregation, creating a deeply divided social world, can cause intergroup disparities across neighborhoods in a metropolitan area (Galster 1992). Separate informal networks and formal institutions serving the less-educated community may have fewer human resources. Thus, they will offer inferior options for the development of human capital (Lim 2010). Based on the context of human capital, recent studies pointed out that segregation between the highly educated and the less educated increased dramatically in the late twentieth century and has had crucial social consequences (Orfield and Lee 2005; Domina 2006; O'Nions 2010). Even though educational segregation is perceived as one of the most serious problems in the metropolitan area, a number of previous empirical studies on residential segregation have still focused on racial, ethnic, and income segregation. In this context, our empirical findings contribute by providing information on residential segregation according to educational attainment. This can inform future research on mechanisms of residential segregation in terms of educational attainment or human capital in the context of growing concerns about educational segregation in recent urban and regional studies. Even though we have not attempted to explain another dimension of residential segregation, such as income, occupation, and housing, our findings can contribute toward providing a valuable policy lesson to lessen uneven patterns in neighborhood and community development.

Moreover, educational segregation may be systematically linked to various dimensions of residential segregation, especially in terms of the limited access to the social networks that link individuals to opportunities. In this context, our discussion of the geography of educational segregation can shed light on the establishment of comprehensive regional policies to lessen socioeconomic inequalities across neighborhoods and enhance social integration. Particularly, a polarized educational geography, as discussed in this paper, may contribute to increasingly uneven patterns in neighborhood and community development. This is because the spatial dimension of human capital segregation affects individual performance through the effect of neighborhood externalities. From this perspective, our discussions on the geography of educational segregation should be expanded to form a more comprehensive approach to understanding inequality in the socioeconomic status of neighborhoods and thereby enhancing our understanding of neighborhood residential segregation, its consequences, the behavioral decisions that underpin it, and the policy options for effectively altering it.

References

Acemoglu, D. 1996. A microfoundation for social increasing returns in human capital accumulation. *Quarterly Journal of Economics* 111: 779–804.

Anselin, L. 1995. Local indicators of spatial association-LISA. *Geographical Analysis* 27: 93–115.

Apparicio, P., J.C. Martori, A.L. Pearson, É. Fournier, and D. Apparicio. 2013. An open-source software for calculating indices of urban residential segregation. *Social Science Computer Review* 32: 117–128.

Audretsch, D., and M.P. Feldman. 2003. Knowledge spillovers and the geography of innovation. In *Handbook of urban and regional economics,* ed. J.V. Henderson and J. Thisse, 2713–2739. Amsterdam: North Holland Publishing.

Azariadis, C., and A. Drazen. 1990. Threshold externalities in economic development. *Quarterly Journal of Economics* 105: 501–526.

Bayer, P., and R. McMillan. 2012. Tiebout sorting and neighborhood segregation. *Journal of Public Economics* 96: 1129–1143.

Bayer, P., H. Fang, and R. McMillan. 2014. Separate when equal? Racial inequality and residential segregation. *Journal of Urban Economics* 82: 32–48.

Benabou, R. 1993. Working of a city: Location, education, and production. *Quarterly Journal of Economics* 108: 619–652.
Brooks-Gunn, J., G.J. Duncan, P.K. Klebanov, and N. Sealand. 1993. Do neighborhoods influence child and adolescent development? *American Journal of Sociology* 99: 353–395.
Cliff, A., and J.K. Ord. 1981. *Spatial processes: Models and applications*. London: Pion.
Crane, J. 1991. The epidemic theory of ghettos and neighborhood effects on dropping out and teenage childbearing. *American Journal of Sociology* 96: 1226–1259.
Cutler, D.M., E.L. Glaeser, and J.L. Vigdor. 1999. The rise and decline of the American ghetto. *Journal of Political Economy* 107: 455–506.
Domina, T. 2006. Brain drain and brain gain: Rising educational segregation in the United States, 1940–2000. *City and Community* 5: 387–407.
Duncan, O.D., and B. Duncan. 1955. A methodological analysis of segregation indexes. *American Sociological Review* 20: 210–217.
Duncan, G.J., and S.W. Raudenbush. 1999. Assessing the effects of context in studies of child and youth development. *Educational Psychologist* 34: 29–41.
Durlauf, S.N. 2004. Neighborhood effects. In *Handbook of regional and urban economics*, ed. J.V. Henderson and J.F. Thisse, 2173–2242. Amsterdam: Elsevier.
Galster, G.C. 1992. A cumulative causation model for the underclass: Implications for urban economic development policy. In *The metropolis in black and white: Place, power and polarization*, ed. E.W. Hill, 190–215. New Brunswick: Center for Urban Policy Research, Rutgers University.
Galster, G.C. 2001. On the nature of neighborhood. *Urban Studies* 38: 2111–2124.
Galster, G.C., and S.P. Killen. 1995. The geography of metropolitan opportunity: A recent reconnaissance and conceptual framework. *Housing Policy Debate* 6: 7–43.
Glaeser, E.L. 1999. Learning in cities. *Journal of Urban Economics* 46: 254–277.
Gordon, I., and V. Monastiriotis. 2006. Urban size, spatial segregation and inequality in educational outcomes. *Urban Studies* 43: 213–236.
Grigoryeva, A., and M. Ruef. 2015. The historical demography of racial segregation. *American Sociology Review* 80: 814–842.
Hårsman, B., and J.M. Quigley. 1995. The spatial segregation of ethnic and demographic groups: Comparative evidence from Stockholm and San Francisco. *Journal of Urban Economics* 37: 1–16.
James, D.R., and K.E. Taeuber. 1985. Measures of segregation. *Sociological Methodology* 14: 1–32.
Jovanovic, B., and R. Rob. 1989. The growth and diffusion of knowledge. *Review of Economic Studies* 56: 569–582.
Leventhal, T., and J. Brooks-Gunn. 2000. The neighborhoods they lived in: The effect of neighborhood residence on child and adolescent outcomes. *Psychological Bulletin* 126: 309–337.
Lim, U. 2008. The spatial dimension of human capital segregation: An empirical investigation for Seoul, Korea. *Social Science Journal* 45: 682–690.
Lim, U. 2010. Do birds of a feather flock together? Exploring the geography of educational segregation in Seoul, Korea. *Studies in Regional Science* 40: 479–494.
Lucas, R.E. 1988. On the mechanics of economic development. *Journal of Monetary Economics* 22: 3–42.
Massey, D.S. 1996. The age of extremes: Concentrated affluence and poverty in the twenty-first century. *Demography* 33: 395–412.
Massey, D.S., and N.A. Denton. 1988. The dimensions of residential segregation. *Social Forces* 67: 281–315.
Massey, D.S., and N.A. Denton. 1993. *American apartheid: Segregation and the making of the underclass*. Cambridge, MA: Harvard University Press.
Mayer, S.E., and C. Jencks. 1989. Growing up in poor neighborhoods: How much does it matter? *Science* 243: 1441–1445.

Moretti, E. 2004. Human capital externalities in cities. In *Handbook of regional and urban economics,* ed. J.V. Henderson and J.F. Thisse, 2243–2291. Amsterdam: Elsevier.
O'Nions, H. 2010. Different and unequal: The educational segregation of Roma pupils in Europe. *Intercultural Education* 21: 1–13.
OECD. 2014. *Education at a glance 2014: OECD Indicators.* Paris: OECD Publishing.
Orfield, G., and C. Lee. 2005. *Why segregation matters: poverty and educational inequality,* The civil rights project, series K-12 integration and diversity. Cambridge: Harvard University. http://escholarship.org/uc/item/4xr8z4wb. Accessed 16 Sept 2015.
Quercia, R.G., and G.C. Galster. 2000. Threshold effects and neighborhood change. *Journal of Planning Education and Research* 20: 146–162.
Ramos, R., J. Surinach, and M. Artis. 2010. Human capital spillovers, productivity and regional convergence in Spain. *Papers in Regional Science* 89: 435–447.
Rauch, J.E. 1993. Productivity gains from geographical concentration of human capital: Evidence from the cities. *Journal of Urban Economics* 34: 380–400.
Reardon, S.F., and G. Firebaugh. 2002. Measures of multigroup segregation. *Sociological Methodology* 32: 33–67.
Reardon, S.F., and D. O'Sullivan. 2004. Measures of spatial segregation. *Sociological Methodology* 34: 121–162.
Reardon, S.F., S.A. Matthews, D. O'Sullivan, B.A. Lee, G. Firebaugh, C.R. Farrell, and K. Bischoff. 2008. The geographic scale of metropolitan racial segregation. *Demography* 45: 489–514.
Rosenthal, S.S., and W.C. Strange. 2008. The attenuation of human capital spillovers. *Journal of Urban Economics* 64: 373–389.
Sampson, R.J. 2009. Racial stratification and the durable tangle of neighborhood inequality. *Annals of the American Academy of Political and Social Science* 621: 260–280.
Sampson, R.J., J. Morenoff, and T. Gannon-Rowley. 2002. Assessing "neighborhood effects": Social processes and new directions in research. *Annals of Review of Sociology* 28: 443–478.
Schlitte, F. 2012. Local human capital, segregation by skill, and skill-specific employment growth. *Papers in Regional Science* 91: 85–106.
Schultz, T.P. 1988. Education investment and returns. In *Handbook of development economics,* ed. H. Chenery, and T.N. Srinivasan, 543–630. Amsterdam: North-Holland.
Smith, N. 1991. *Uneven development: Nature, capital, and the production of space.* Oxford: Blackwell.
Swanstrom, T., P. Dreier, and J. Mollenkopf. 2002. Economic inequality and public policy: The power of place. *City and Community* 1: 349–372.
Timberlake, J.M. 2002. Separate, but how unequal? Ethnic residential stratification, 1980 to 1990. *City and Community* 1: 251–266.
Upton, G., and B. Fingleton. 1985. *Spatial data analysis by example.* New York: Wiley.
Vartanian, T.P., and P.M. Gleason. 1985. Do neighborhood conditions affect high school dropout and college graduation rates? *Journal of Socio-Economics* 28: 21–41.
Wilson, W.J. 1987. *The truly disadvantaged: The inner city, the underclass, and public policy.* Chicago: University of Chicago Press.
Wong, D.W. 2005. Formulating a general spatial segregation measure. *Professional Geographer* 57: 285–294.

Labor Market Distortion with Discouraged Worker Effects in Korea

Jaewon Lim

Abstract This paper examines skill mismatch problems in regional labor markets due to discouraged worker effects in South Korea. In Seoul metropolitan areas (SMAs), supply surplus for highly educated workforce among the youth is evident and causes serious distortion not only in SMA but also in national labor market. On the contrary, the shortage of low-skilled workers in Korea due to overinvestment in human capital forces out establishments to other countries with cheaper labor costs. Proposed regression models in this study specify how the share of discouraged workers in a regional labor market can be determined by various factors such as demographic structures, labor market conditions, and migration pattern with regional and/or temporal fixed effects. The regional-temporal fixed effect model found to be the most important factor in explaining how these factors determine the relative stock of discouraged workers in a region. Among the variables describing regional labor market conditions, only two, labor force participation rate of young cohort (aged between 15 and 29) and that of highly educated population (college graduate or higher), directly influenced regional labor market distortion by determining the regional stock of discouraged workers. With more active participation of these groups in a labor market, the overall discouraged worker effect in a region reduces, leading to enhanced labor market efficiency. Migration pattern does not play any role on regional discouraged workers. This indicates the limited role of interregional migration on factor price equalization among regional labor markets in Korea.

Keywords Regional labor market • Spatial mismatch • Skilled workforce • Labor force participation

---

J. Lim (✉)
School of Public Policy & Leadership, Greenspun College of Urban Affairs, University of Nevada, Las Vegas, NV, USA
e-mail: jaewon.lim@unlv.edu

E. Kim, B.H.S. Kim (eds.), *Quantitative Regional Economic and Environmental Analysis for Sustainability in Korea*, New Frontiers in Regional Science: Asian Perspectives 25, DOI 10.1007/978-981-10-0300-4_3

1 Introduction

Unemployment and underemployment among young labor force emerges as one of the main causes that distorts the overall labor market efficiency in advanced economies. Among other factors, unstable macroeconomic condition, rapidly aging population, and skill mismatch problems in labor market directly affect the unemployment and underemployment of young adult. First, fluctuation in macroeconomic condition, like great recession (2007–2009) in the USA and subsequent economic downturns among other advanced countries, causes the structural shift in labor market demand, with relatively inelastic supply of labor with respect to external economic shock. Moreover, even with rapid economic recovery in recent years, expanded job opportunities are still limited to and available only for certain groups of labor force with specific skillsets. Young labor force with the lack of work experience has mostly suffered from the unemployment during recession and underemployment during recent recovery. Especially, many young and highly skilled workers have lost opportunities to enter official labor market as full-time employees for jobs that require relevant skill level. This forgone opportunity delays their successful transition to stable labor market status in contrast to underemployment, unemployment, and, more importantly, forced dropout from labor force (though this can be counted as voluntary dropouts from labor force due to economic conditions). However, this can be a temporary distortion in labor market during recession in a country like the USA with highly mobile young labor force and regionally diversified labor market conditions. In other words, young and skilled workforce can and tends to broaden job search nationwide. This process contributes to factor price equalization in labor market across the USA, resolving labor market distortion in the long run. In contrast, young and highly mobile workforce of a country like Republic of Korea (Korea, hereafter), with highly concentrated economy in its economic heart, Seoul metropolitan area (SMA), tends to stuck, and their mobility toward factor price equalization is largely limited. Second, aging population in many advanced economies translates into (1) delayed retirement and (2) contraction in consumption. Hence, even with the smaller share of young labor force, it is expected to be tougher for them to enter full employment, mainly due to demographic transition. Compared to the macroeconomic fluctuation, this will last longer causing chronic distortion in national and local labor markets. Last but not least, skill mismatch is another factor causing unemployment and underemployment of young labor force. One good example is Korea that suffers from bipolarized labor market distortion. On the one hand, supply surplus of highly educated young workforce is mainly concentrated in the SMA of Korea even with much higher competitions than the rest of Korea (ROK). With limited demand for highly skilled workforce, a growing portion of highly educated young workforce voluntarily drops out of labor force. This discouraged worker effect in Korea is mainly driven by overinvestment in human capital, causing serious distortion in national and regional labor markets. In addition, the supply shortage of low-skilled workers in ROK due to voluntary dropouts associated with overinvestment in human capital forces out establishments to other countries where they can easily access more abundant cheaper labor supply. Theoretically, supply shortage of low-skilled labor force in ROK can be resolved by filtering down process, the inflow of young highly skilled labor force surplus from SMA to the employment opportunities in ROK. This process further aggravates labor underutilization but helps maintain overall efficiency in national labor market. However, in reality this rarely happens for two reasons: (1) young labor force migration from the ROK to SMA is very rare, and (2) highly educated young workforce in Korea is very reluctant to be underemployed. In the long run, the limited filtering down across regional labor markets in Korea may proliferate discouraged workers, resulting in overall distortion in labor market. Labor market in the USA is relatively free from these issues with the diversified regional labor market providing a variety of destination choices for younger workforce migrants. The skill mismatch problem caused by overinvestment human capital in Korea significantly distorts labor market among regions, and it is very challenging to resolve with policy instruments.

This paper examines labor market distortion due to the skill mismatch problems and limited mobility across 15 regions[1] in Korea. A set of proposed panel data regression models describe how discouraged workers in regional labor markets can be estimated by regional demographic structure, regional labor market conditions, and migration efficiency of a region.

The following section summarizes the previous studies on the topics of discouraged worker effects and underemployment, human capital investment for regional economic development, and migration of high-skilled workers. In Sect. 3, data and analytical approach for empirical model is introduced. Section 4 shares empirical results and major findings, followed by conclusion and further discussion on policy implications in Sect. 5.

2 Theoretical Review

2.1 Discouraged Worker Effects and Underemployment

According to the Bureau of Labor Statistics (BLS) in the USA, marginally attached individuals are not counted as unemployed since they are not part of a labor force due to inactive job search efforts for the very previous 4-week period prior to survey. However, they are still available for work and willing to work and have searched for jobs in the prior 12-month period. Discouraged workers are defined as a part of those who are marginally attached to labor force. The reason for quitting

[1]The fifteen regions include seven metropolitan areas, Seoul, Incheon, Busan, Daegu, Gwangju, Daejeon, and Ulsan, and eight provinces (excluding the seven metropolitan areas). The eight provinces are Gyeonggi, Gangwon, Chungbuk, Chungnam, Jeonbuk, Jeonnam, Gyeongbuk, and Gyeongnam.

job search distinguishes discouraged worker from all the other types of marginally attached individuals. Discouraged workers believe either there are no jobs available for them or they would not qualify for none of the available jobs. Among other types of marginally attached workers, part-time workers for economic reasons serve as a major source for labor underutilization. BLS releases annualized alternative measures of labor underutilization among US states taking into account the presence of marginally attached workers. Specifically, three alternative measures (U-4, U-5, and U-6) are more inclusive and serve as comprehensive measures for labor underutilization (Haugen 2009). In contrast to the official unemployment rate measure (U-3[2]), these three alternative measures count the following two groups, (a) the part-time workers for economic reasons and (b) those who quit job searching within the last 4 weeks because they believe either there are no jobs available for them or they would not qualify for none of the available jobs, as part of unemployed labor force. During the recent recession, diminishing labor market efficiency captured by the outward shift of Beveridge curve[3] one occurred over a short period (less than 1 year), as compared to the roughly 8 years it took for the Beveridge curve to shift out in the recession of the 1970s (Ghayad and Dickens 2012). Among others, Kroft et al. (2014) found that the increasing long-term unemployment during the most recent recession was causing labor market inefficiencies associated with the negative duration dependence in the job-finding rate. With empirical evidence, David et al. (2015) showed that additional months out of the labor force have a negative effect on the probability of transition into employment. Their empirical work proved how the delayed processing time for Social Security Disability Insurance (SSDI) affected the employment and earnings of SSDI applicants in subsequent years. Earlier studies by Okun et al. (1973) and Blanchard and Diamond (1994) formulated the models of queuing and ranking to explain negative duration in the job-finding rate for nonparticipants in the labor force. These models put marginally attached workers at the end of the queue for job-finding activities. More recently, Barr and Turner (2013) found that some adult workers returned to school and/or job training in part attributed to the negative duration dependence in job-finding rate for nonparticipants. For any reason, lower-than-expected job-finding rate of nonparticipants during and in the aftermath of great recession is known to be the primary source of increasing long-term unemployment. Elsby et al. (2013) noted that the variation in labor force participation represents one-third of the cyclical variation in the unemployment rate. Hence, ignoring the fluctuation in labor force participation leads to substantial underestimates of the overall unemployment rate during the great recession. As a consequence, flows between "unemployed" (U) and "not-in-labor force" (NILF) provide the basis for

[2]Official unemployment rate (U-3) is defined as a ratio of the number of unemployed to the total labor force, where the total labor force is composed of unemployed and employed.

[3]Beveridge curves describe the negative relation between unemployment rates and job vacancy rates in labor markets. Outward shift of Beveridge curve indicates increasing inefficiency due to potential mismatch between those unemployed and job openings.

understanding long-term unemployment and the movement of labor market efficiency indices. Coibion and Gorodnichenko (2013) summarized the factors raising persistency in unemployment during the recession periods in the USA, including rising cyclicality in long-term unemployment, rising cyclicality in disability claims, missing disinflation, and lower regional convergence after the downturn. Among these four factors, lower regional convergence draws special attention since this may presage a diversified recovery path by region with the emerging importance of labor force participation in the aftermath of great recession. Regional diversity in labor market conditions, considering the potential role of rising long-term unemployment, can be analyzed from both the demand and supply sides. Some regions may suffer from the prolonged shortage of labor demand, while other regions may have much quicker increases in labor demand. One of the frequently found supply-side issues is the negative duration dependence in the transition from the unemployed to the employed. Populations classified as NILF (including all marginally attached workers) are another source for supply-side problems since this population may further slow down the recovery due to the higher negative dependence with a longer period of unemployment. The issue here is that the distinction between "unemployed" and "not-in-labor force" is not always clear.

Much economic analysis is based on the "unemployed" (U) versus "not-in-labor force" (NILF) distinction, with those classified as U often modeled as being engaged in optimal search behavior and those classified as NILF as being engaged in household production. Most models of labor market behavior treat those classified as U as being willing to work at the market wage (and thus at an interior solution in terms of desired hours of work), while those classified as NILF are modeled as being at a corner solution, thus requiring a higher offered wage to entice them into the labor market. Finegan (1978) and Sorrentino (1995) attempted to solve the issues on how to treat the non-searching unemployed for labor statistics at national level. Regarding labor supply estimates in the USA, Finegan (1978) points out the potential problems with a job-seeking test due to the discouraged workers causing losses for society and for themselves. Much research verified the potential distortion of labor statistics due to the discouraged worker effect, represented by the decreasing number of unemployed actively searching jobs facing adverse economic conditions (Blundell and Bond 1998; Tachibanaki and Sakurai 1991; Kuch and Sharur 1978; Ondeck 1978). Previous empirical studies tested the vague distinction between U and NILF focusing on job-searching behaviors (Clark et al. 1979; Flinn and Heckman 1983; Gönül 1992; Jones and Riddell 1999) with the comparison of the two groups' transition probabilities from the unemployed to the employed. Within advanced economies in Europe and North America, there is significant empirical evidence about the presence of spatial differentials in unemployment rates (see, for example, López-Bazo et al. 2002). Elhorst (2000) recommends analyzing unemployment and labor underutilization in a regional context for the following reasons: the varying regional differences within countries, the lack of theory for the existing regional disparities in macroeconomics, and the inefficiency caused by regional unemployment disparities. Most of the empirical studies in Korea mainly focus on the nationwide temporal transition, rather than on the spatial aspects of labor market conditions. However, both supply- and demand-side conditions vary by subnational regions depending on the regional economic structure and the regional efforts to improve labor market conditions. Regional labor market conditions can be more accurately described by considering labor underutilization caused by discouraged workers due to economic conditions. Since young workforce even with a higher level of human capital suffers from a higher barrier to entry for labor market, it is critical to understand how the regional distribution of young and/or highly educated workforce contributes to the accumulation of discouraged workers among Korean regions.

2.2 Human Capital Investment for Regional Economic Development

Endogenous growth models attribute long-run economic growth to the accumulation of knowledge and of human capital (see Romer 1989; Lucas 1988). Bal and Nijkamp (1997) describe human capital investment as endogenous driving force further extending neoclassical Solow-Swan model to new growth theory. Endogenous technological change can be initiated and advanced by both public and private investment for human capital and R&D activates (Nijkamp and Poot 1998). Factor mobility in an advanced and open economy plays a critical role for technological advancement through spatial interaction. Other forms of spatial interaction include knowledge diffusion and trade. Factor mobility foresees factor price equalization through the flow of human capital from a region with surplus (at a lower price) to a region with shortage (at a higher price). Instead, the exact reverse flow, brain drain, is more evident in reality at international setting (Lucas 1988). Even at the subnational level, "cumulative causation" is more evident rather than regional convergence (Van Dijk et al. 1989). This is largely due to asymmetric information, imperfect labor markets, and adjustment cost associated with migration (Barro and Sala-i-Martin 1995; Gordon and Bovenberg 1996). Reallocation of human capital and accompanying diffusion of knowledge is a critical factor for economic growth of a region that may reduce regional disparities. However, certain regions are not prepared to adopt diffused knowledge including human capital mainly due to a local condition (Nijkamp and Poot 1998). Consequently, relevant policy instruments for regional knowledge accumulation may greatly vary depending on local conditions. Public investment cannot solely deal with regional knowledge accumulation; rather it should be combined with private investments focusing on region-specific attribute of technology and innovation structure. Cross-national comparison by Kubo and Kim (1996) clearly identified a robust complementary association between human capital accumulation and technology adoption and in their case study for Korea and Japan.

2.3 Migration of High-Skilled Workers

Empirical studies on migration and education indicate that highly educated are the most mobile among the US population and they are less reluctant to migrate repeatedly and willing to move long distances (Basker 2002; Kodrzycki 2001). Mobility of highly educated may create a concentration of human capital in certain regions attracting more human capital, and this serves as a driving force for regional economic growth (Mathur 1999). Waldorf (2009) found that the highly educated tend to locate in large-scale metropolitan areas forming agglomeration economies (Waldorf 2009). Others pointed out that the abundant managerial and professional jobs worked as pulling factors for the highly educated (Costa and Kahn 2000; Schachter et al. 2003). Among many potential pulling factors, regional human capital stock pulls highly educated migrants and discouraged out-migrants of the highly educated (Gottlieb and Joseph 2006). As a consequence, metropolitan areas which fail to accumulate the critical mass of highly educated are more likely to fall behind in knowledge-based economy (Waldorf 2009). As Kaldor (1970) discussed, cumulative causation in regional economic growth strengthens the reinforcing linkages between specialization and competitive advantage at regional scale. Regional disparity further aggravates due to the spatial concentration of highly educated workforce. Waldorf (2009) analyzed the stock-flow relationships of the highly educated utilizing a series of regression models that control for the degree of rurality, spatial (geographical) trends, and other sociodemographic characteristics. The empirical results confirm that the migration pattern of highly educated population aggravates disparities among the regions with different levels of human capital endowment, leading to "divergence of human capital levels" across space as defined by Berry and Glaeser (2005). Earlier migration literature indicates that highly educated and specialized human capital tends to migrate longer distance to exploit larger financial rewards in a more spatially extensive labor market (Sjaastad 1962; Schwartz 1973; Greenwood 1975). This finding reveals that regional development policies with human capital investment without the proper local employment opportunities retain highly educated human capital in the region. Faggian et al. (2006, 2007) classified the sequential migration of higher education institution (HEI) graduates into five types, namely, repeat migrants, return migrants, late migrants, university stayers, and nonmigrants. Two major findings from their proposed model indicate that, first, university graduates are highly mobile and most students do not enter employment in the same area and, second, knowledge spillover effects of universities are very limited and largely determined by the strength of local economy as a whole, rather than by the quality of university.

3 Data and Model Specification

This paper specifies estimation models for discouraged worker effects among 15 Korean regional labor markets using annual labor market data from 2003 to 2014. Subnational regions for this study include seven metropolitan areas and eight

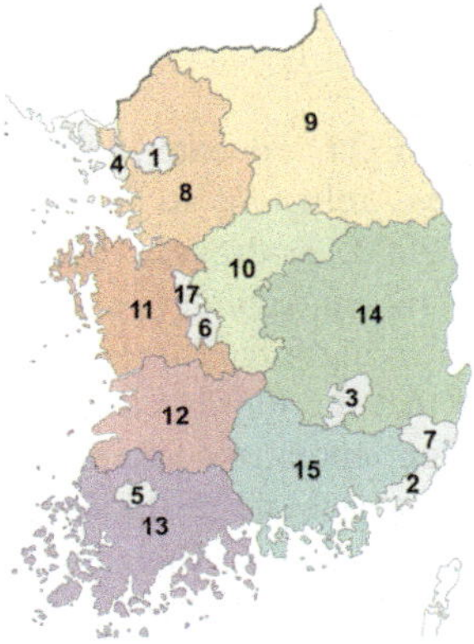

Region ID	Type	Name	Region ID	Type	Name
1	Metro	Seoul	8	Province	Gyeonggi
2	Metro	Busan	9	Province	Gangwon
3	Metro	Daegu	10	Province	Chungbuk
4	Metro	Incheon	11	Province	Chungnam
5	Metro	Gwangju	12	Province	Jeonbuk
6	Metro	Daejeon	13	Province	Jeonnam
7	Metro	Ulsan	14	Province	Gyungbuk
			15	Province	Gyungnam

[a]Newly created metropolitan area, Sejong city (17) and Jeju Province (not shown on map) are omitted due to the lack of data for analysis.

Fig. 1 15 subnational regions[a] in Korea

provinces (see Fig. 1). Data source for labor statistics is the Korean *Ministry of Employment and Labor* (MOEL)[4], and it produces four major types of survey data: (1) Labor Force Survey at Establishments, (2) Labor Cost of Enterprise Survey, (3) Labor Demand Survey, and (4) Survey on Labor Conditions by Type of Employment.

Variables for model specification are collected under panel data structure at the regional level for a 15-year period (2003–2014) on annual basis. All the variables fall into one of the following three categories: (1) demographic structure (age and education attainment), (2) labor market conditions by age and education attainment (reasons for NILF, labor force participation, unemployment rate, job openings, and job seekers), and (3) migration efficiency (region-specific migration efficiency). Descriptive statistics of variables is outlined in Table 1.

Under panel data structure, there are a total of 180 observations over a 12-year period for 15 regions. For model specification, this paper specifies four types of linear regression models for panel data: pooled OLS model, two fixed effect models, and random effect model, as shown in Eqs. (1), (2), (3), and (4), respectively.

- Pooled OLS Regression

$$y_{it} = x_{it}'\beta + \varepsilon_{it} \tag{1}$$

where

y = share of not-in-labor force (NILF) population for economic reasons out of the total population aged 15 years or older
x = set of explanatory variables (described in Table 1)
β = estimated coefficients for explanatory variables

[4]For more detail on the Korean labor statistics, visit the following link, http://laborstat.moel.go.kr/

Table 1 Descriptive statistics of variables

Variable	Description	*N*	Mean	Median	Std. Dev	Min	Max
SH_POP1564	Share of Population aged 15–64 out of Total Population	180	.858130	.865666	.044522	.742455	.936881
SH_POP1529	Share of Population aged 15–29 out of Total Population	180	.239380	.235874	.030155	.173791	.320037
SH_POP2529	Share of Population aged 25–29 out of Total Population	180	.086564	.085794	.016706	.042924	.120344
SH_POP65h	Share of Population aged 65 and older out of Total Population	180	.141870	.134334	.044522	.063119	.257545
SH_POP_AD_h	Share of Population with Associate or Higher Degree	180	.265172	.259477	.065275	.143047	.436788
SH_POP_CG	Share of Population with College Graduate or Higher	180	.174851	.160139	.054920	.080506	.339829
SH_TLF_AD_h	Share of Total Labor Force with Associate or Higher Degree	180	.340588	.331181	.082518	.184369	.532294
SH_TLF_CG	Share of Total Labor Force with College Graduate or Higher	180	.225262	.207775	.068724	.103051	.408608
SH_NILF_ER_POP	Share of NILF for Economic Reason out of Total Population aged 15–64	180	.050812	.051513	.007391	.030143	.067354
SH_NILF_AOR_POP	Share of NILF for All Other Reasons out of Total Population aged 15–64	180	.106563	.104534	.015717	.071689	.146942
LFPR_1564	Labor Force Participation Rate of Population aged 15–64	180	.654914	.653305	.022842	.606484	.700967
LFPR_1529	Labor Force Participation Rate of Population aged 15–29	180	.427746	.422984	.045548	.314478	.539947
LFPR_2529	Labor Force Participation Rate of Population aged 25–29	180	.710134	.715385	.041393	.582000	.784259
LFPR_LTES	Labor Force Participation Rate of Population with Less Than Elem School	180	.418093	.398735	.080387	.284742	.617143
LFPR_MSG	Labor Force Participation Rate of Population with Middle School Degree	180	.428455	.424687	.052948	.308029	.550000
LFPR_HSG	Labor Force Participation Rate of Population with High School Diploma	180	.644814	.645509	.035739	.572222	.720403
LFPR_AD	Labor Force Participation Rate of Population with Associate Degree	180	.776245	.778808	.026234	.702500	.859794
LFPR_CG	Labor Force Participation Rate of Population with College or Higher Degree	180	.785087	.781359	.027335	.724955	.864626

(continued)

Table 1 (continued)

Variable	Description	*N*	Mean	Median	Std. Dev	Min	Max
LFPR_AD_h	Labor Force Participation Rate of Population with Associate or Higher Degree	180	.781678	.776933	.023021	.732468	.847541
UEMR_1564	Unemployment Rate of Labor Force aged 15–64	180	.029833	.028877	.008584	.013802	.047570
UEMR_1529	Unemployment Rate of Labor Force aged 15–29	180	.070019	.069895	.013926	.036254	.110334
UEMR_2529	Unemployment Rate of Labor Force aged 25–29	180	.058221	.056249	.015708	.025276	.102699
RD2S_all	Ratio of Job Openings to Job Seekers for all types of full time jobs	180	.570300	.530153	.209067	.258243	1.176394
RJAS_all	Ratio of New Hires to Job Seekers for all types of full time jobs	180	.251167	.250988	.048325	.126220	.381227
MIGER_all	Migration Efficiency Rate [= (Net Migration) / (sum of In- & Out Migration)] for all age	180	−.014053	−.011297	.059971	−.247958	.153778
MIGER_1519	Migration Efficiency Rate [= (Net Migration) / (sum of In- & Out Migration)] for aged 15–19	180	−.039573	−.042232	.082681	−.333986	.224324
MIGER_2024	Migration Efficiency Rate [= (Net Migration) / (sum of In- & Out Migration)] for aged 20–24	180	−.066689	−.074586	.114014	−.358047	.240078
MIGER_2529	Migration Efficiency Rate [= (Net Migration) / (sum of In- & Out Migration)] for aged 25–29	180	−.043707	−.041521	.093134	−.267547	.217834

Source: author's own calculation

i = subnational study region in Korea ($i = 1, 2, \ldots, 15$)
t = year for study period ($t = 2003, 2004, \ldots, 2014$)
ε_{it} = error term

- Regional Fixed Effect Model

$$y_{it} = x'_{it}\beta + \alpha_i + \varepsilon_{it} \tag{2}$$

where

α_i = regional fixed effect term (region-specific constant term)
All other terms are the same as in Eq. (1)

- Regional-Temporal Fixed Effect Model

$$y_{it} = x'_{it}\beta + \alpha_i + \gamma_t + \varepsilon_{it} \tag{3}$$

where

γ_t = temporal fixed effect term (time-specific constant term)
All other terms are the same as in Eqs. (1) and (2)

- Random Effect Model

$$y_{it} = x'_{it}\beta + \alpha + u_i + \varepsilon_{it} \tag{4}$$

where

u_i = region-specific random element
All other terms are the same as in Eqs. (1), (2), and (3)

Proposed models are designed to estimate the share of "NILF" due to economic conditions out of the total population aged 15 and over. This group of population would have attached to labor force unless there is no negative economic condition. Though classified as marginally attached workers, this group of NILF population does not fall into labor force or unemployed. For this reason, official unemployment rate[5] tends to underestimate labor underutilization in Korea. This is also a well-known issue in labor statistics in many countries including the USA; however, specific reasons for voluntary dropouts from labor force are usually aggregated. From the Korean labor statistics, reasons for not participating labor force among population aged 15 and over are summarized in six categories: child care, home-maker, studying, retired, sick, or other reason. The last category, "other reason," includes the population who are out of labor force due to economic condition. The time series data is available at national level. But at regional level, the "other

[5]Official unemployment rate in Korea is the share of unemployed out of the total labor force, which is composed of employed and unemployed. Those who voluntarily drop from labor force due to economic condition are not taken into account.

reason" is aggregated with "retired" and "sick" categories. For this reason, it was required to decompose under the following two assumptions. First, the share of "retired" in a region is proportional to the share of population 65 years and older. Second, the share of "sick" in a region follows the same share at national level. So, the dependent variable in this study represents the share of discouraged workers (currently not-in-labor force) due to economic condition among population 15 years and older.

To estimate the regionally varying discouraged worker effects, the proposed models employ a set of explanatory variables on regional demographic structure (the share of young age cohorts and share of highly educated population), another set of variables on labor market conditions (job opening to job seeker ratio, labor force participation rate, and unemployment rate of the young age cohort and/or of the highly educated), and, finally, a variable showing regional migration (migration efficiency rate of a region).

4 Estimation Results

Estimation results are summarized for four model specifications in Table 2. Dependent variable is "*SH_NILF_EF_POP*" (the share of NILF for economic reason of the total population aged 15–64), and this is observed among 15 subnational Korean regions from 2003 to 2014, annually. Among the four models, regional-temporal fixed effect model (3) is the best fit under panel structure. Model (3) considers both regional and temporal fixed effects for panel data.

A series of test statistics to identify the best-fit model is performed among the four models. The test statistics results are summarized in Table 3. The first three tests are designed to compare models (1), (2), and (4), and the results indicate that fixed effect model (2) is superior to random effect model (4). Compared to pooled OLS model (1), *F-test* supports fixed effect model (2) with significant individual effect from subnational regions in Korea. Consequently, fixed effect model (2) with regional fixed effect is the best model specification among the three models. For the following three test statistics listed in Table 3, another fixed effect model (3) with two fixed effects, namely, regional and temporal, is tested against all three other models. Hausman tests support for both fixed effect models, (2) and (3), against random effect model, (4). Finally, two fixed effect models are compared against using *F-test* statistics for individual (regional) effects. Test result supports model (3) (regional-temporal fixed effect model) against model (2) (regional fixed effect model). Adjusted R-square of fixed effect model significantly improves from model (2) at 0.4311 to model (3) at 0.6647.

All the coefficients for regional fixed effects are statistically significant as summarized in Table 4 for two fixed effect models (2) and (3). Additionally, regional fixed coefficients of the subnational regions are compared to that of Seoul, and this can be found in the column with heading, "compared to Seoul." Interestingly, when temporal fixed effect is not considered (model (2)), all

Table 2 Estimation result

	(1) Pooled OLS		(2)Fixed effect (Region)		(3) Fixed effect (Region & time)		(4)Random effect	
Variables	Coefficient	*p*-value	Coefficient	*p*-value	Coefficient	*p*-value	Coefficient	*p*-value
(intercept)	0.1904	< 2.22e-16***					0.1642	< 2.22e-16***
RJAS_all_LG	0.0049	0.59856	−0.0227	0.00658**	0.0076	0.33980	−0.0132	0.11211
SH_POP1529	0.0413	0.03721*	−0.0790	0.00422**	0.1136	0.00005 ***	−0.0333	0.18430
LFPR_1529	−0.0814	7.38E-11***	−0.0809	3.30E-06***	−0.0567	1.57E-05 ***	−0.0858	1.01E-07***
UEMR_1529_LG	0.0365	0.37822	0.0455	0.19427	−0.0176	0.32680	0.0634	0.07683 .
LFPR_CG	−0.1494	2.87E-15***	−0.0650	9.51E-05***	−0.0575	3.52E-05 ***	−0.0856	4.36E-07***
UEMR_CG_LG	−0.0334	0.66631	−0.1098	0.04984*	−0.0396	0.31340	−0.1004	0.08604 .
MIGER_all	0.0150	0.04677*	0.0133	0.10241	−0.0017	0.77340	0.0153	0.05462 .
factor(YEAR_2004)					0.0048	2.20E-16***		
factor(YEAR_2005)					0.0123	2.20E-16***		
factor(YEAR_2006)					0.0119	2.20E-16***		
factor(YEAR_2007)					0.0116	2.20E-16***		
factor(YEAR_2008)					0.0112	2.20E-16***		
factor(YEAR_2009)					0.0131	2.20E-16***		
factor(YEAR_2010)					0.0128	2.20E-16***		
factor(YEAR_2011)					0.0161	2.20E-16***		
factor(YEAR_2012)					0.0122	2.20E-16***		
factor(YEAR_2013)					0.0113	2.20E-16***		
factor(YEAR_2014)					0.0101	1.00E 11***		
R-Squared	0.5603		0.4912		0.8139		0.4509	
Adj.HR-Squared	0.5354		0.4311		0.6647		0.4309	
F-statistic	31.3165		21.7885		35.7094		20.1783	

(continued)

Table 2 (continued)

	(1) Pooled OLS		(2)Fixed effect (Region)		(3) Fixed effect (Region & time)		(4)Random effect	
Variables	Coefficient	*p*-value	Coefficient	*p*-value	Coefficient	*p*-value	Coefficient	*p*-value
df	172		158		147		172	
p-value	< 2.22e-16***		< 2.22e-16***		< 2.22e-16***		< 2.22e-16***	

Note: signif. codes: 0 '***' 0.001 '**' 0.01 '*' 0.05 '.' 0.1 ' ' 1
Dependent variable: "*SH_NILF_EF_POP*" (share of NILF for economic reason of total population aged 15–64)
$R = 15$ (# of regions) and $T = 12$ (# of annual periods)
$N = 180$ ($=15 \times 12$)

Table 3 Test statistics

Test	*Pooled (1) vs. Random (4)*	*Fixed (region) (2) vs. Poole (1)*	*Fixed (region) (2) vs. Random (4)*	*Fixed (region) (3) vs. Random (1)*	*Fixed (region) (3) vs. Random (4)*	*Fixed (region) (3) vs. Random (2)*
	LM Test (Breusch-Pagan)	*F-test*	*Hausman-Test*	*F-test*	*Hausman Test*	*F-test*
F		14.553		30.922		23.169
Chi-2	103.490		17.571		52.101	
df	1		7		7	
df1		14		25		11
df2		158		147		147
p-value	< 2.22e-16***	< 2.22e-16***	0.000263***	< 2.22e-16***	5.578e-09***	< 2.22e-16***
Support	*Random (4)*	*Fixed (2)*	*Fixed (2)*	*Fixed (region-time) (3)*	*Fixed (region-time) (3)*	*Fixed (region-time) (3)*

14 regions show lower coefficient (impacts) on the "share of not-in-labor force population for economic reasons" than that for Seoul; however, taking into account both regional and temporal fixed effects in model (3), only seven regions read lower coefficient (impacts) compared to Seoul. This indicates that all other variables being constant, these seven regions have lower share of NILF for economic reasons out of the total population aged between 15 and 64 due to the regional fixed effects. Among six metro areas (MAs) other than Seoul, three MAs, Incheon, Daejeon, and Ulsan, have lower impact, whereas four of eight provinces show lower impact and these are Chungnam, Jeonbuk, Jeonnam, and Gyeongbuk provinces. The biggest gap compared to Seoul was found in Chungnam province, followed by Daejeon metro area. These two regions are the direct beneficiaries from the relocation of executive branch of the Korean government to Sejong city since the Korean government first announced its plan for relocation in 2002. The actual relocation has started since 2012, to Sejong city, next to Daejeon MA, the state capital of Chungnam province.

For the best-fit model (3), additional tests to check the validity of linear model have been made and the results are in Table 5. First, cross-sectional dependence in panel data has been tested using Breusch-Pagan LM test of independence. Null hypothesis of no cross-sectional dependence cannot be rejected for the regional-temporal fixed effect model. Secondly, serial correlation in idiosyncratic errors is tested with Breusch-Godfrey/Wooldridge test and serial correlation is evident. Thirdly, Breusch-Pagan test checks homoscedasticity and indicates the presence of heteroskedasticity.

In order to control the identified serial correlation and heteroskedasticity, Sandwich estimator, a robust covariance matrix estimation method, is employed. More

Table 4 Coefficient for regional fixed effect

ID	Region	(2) Fixed effect (Region)			(3) Fixed effect (Region & Time)		
		Coefficient	Compared to Seoul	*p*-value	Coefficient	Compared Seoul	*p*-value
1	Seoul	0.1687	1.0000	< 2.20E-16***	0.0830	1.0000	1.43E-09***
2	Busan	0.1652	0.9792	< 2.20E-16***	0.0876	1.0557	1.01E-11***
3	Daegu	0.1624	0.9631	< 2.20E-16***	0.0833	1.0029	3.21E-10***
4	Incheon	0.1650	0.9782	< 2.20E-16***	0.0825	0.9941	1.58E-09***
5	Gwangju	0.1680	0.9962	< 2.20E-16***	0.0854	1.0290	5.94E-10***
6	Daejeon	0.1616	0.9582	< 2.20E-16***	0.0772	0.9302	2.38E-08***
7	Ulsan	0.1625	0.9635	< 2.20E-16***	0.0811	0.9768	1.87E-09***
8	Gyeonggi	0.1660	0.9840	< 2.20E-16***	0.0847	1.0208	1.86E-10***
9	Gangwon	0.1631	0.9670	< 2.20E-16***	0.0902	1.0869	4.74E-13***
10	Chungbuk	0.1613	0.9565	< 2.20E-16***	0.0843	1.0154	6.72E-11***
11	Chungnam	0.1510	0.8952	< 2.20E-16***	0.0751	0.9046	6.50E-09***
12	Jeonbuk	0.1545	0.9159	< 2.20E-16***	0.0825	0.9932	4.74E-11***
13	Jeonnam	0.1467	0.8699	< 2.20E-16***	0.0803	0.9666	5.09E-11***
14	Gyungbuk	0.1539	0.9123	< 2.20E-16***	0.0795	0.9575	6.46E-10***
15	Gyungnam	0.1586	0.9403	< 2.20E-16***	0.0832	1.0017	9.34E-11***

specifically, *Arellano* estimator, a heteroskedasticity-consistent covariance estimator, can be used to control serial correlation for fixed effect model. The t-test statistics using *Arellano* estimator still estimates same coefficients, but the t values are different from original coefficients (Table 6).

All the significant coefficients from original model specification are still significant even when using *Arellano* estimator. Test statistics shown in Table 2 for model (3), regional-temporal fixed model, report the heteroskedasticity-consistent coefficients (from Arellano estimator).

The largest effect on regional discouraged worker effect (dependent variable) is from regional demographic structure measured by age (*SH_POP1529*) with statistically significant coefficient of 0.1136. A higher share of young working population in a region leads to a higher discouraged effect in the region. In other words,

Table 5 Coefficient for regional fixed effect

Test	Breusch-Pagan LM test	B-G/W test for serial correlation	Breusch-Paga test
Chi-2	1125.56	40.515	
bp			44.926
df	105	12	21
p-value	0.08366	<2.22e-16***	0.001771**

Table 6 Coefficient for regional fixed effect

	Original coefficient		Heteroskedasticity-consistent coefficient (Arellano)	
Variables	Coefficient	*p*-value	Coefficient	*p*-value
RJAS_all_LG	0.0076	0.27398	0.0076	0.33980
SH_POP1529	0.1136	0.00055***	0.1136	0.00005***
LFPR_1529	−0.0567	1.36E-05***	−0.0567	1.57E-05***
UEMR_1529_LG	−0.0176	0.45566	−0.0176	0.32680
LFPR_CG	−0.0575	2.96E-07***	−0.0575	3.52E-05***
UEMR_CG_LG	−0.0396	0.27245	−0.0396	0.31340
MIGER_all	−0.0017	0.75860	−0.0017	0.77340
factor(YEAR_2004)	0.0048	3.26E-07***	0.0048	2.20E-16***
factor(YEAR_2005)	0.0123	< .22e-16***	0.0123	2.20E-16***
factor(YEAR_2006)	0.0119	< .22e-16***	0.0119	2.20E-16***
factor(YEAR_2007)	0.0116	< .22e-16***	0.0116	2.20E-16***
factor(YEAR_2008)	0.0112	6.31E-15***	0.0112	2.20E-16***
factor(YEAR_2009)	0.0131	< 2.22e-16***	0.0131	2.20E-16***
factor(YEAR_2010)	0.0128	1.79E-14***	0.0128	2.20E-16***
factor(YEAR_2011)	0.0161	< .22e-16***	0.0161	2.20E-16***
factor(YEAR_2012)	0.0122	3.43E-11***	0.0122	2.20E-16***
factor(YEAR_2013)	0.0113	2.68E-09***	0.0113	2.20E-16***
factor(YEAR_014)	0.0101	3.87E-07***	0.0101	1.00E-11***

Note: signif. codes: 0 '***' 0.001 '**' 0.01 '*' 0.05 '.' 0.1 ' ' 1
Dependent variable: "*SH_NILF_EF_POP*" (share of NILF for economic reason of total population aged 15–64)
$R = 15$ (# of regions) and $T = 12$ (# of annual periods)
$N = 180$ ($= 15 \times 12$)

10 % increase in population aged between 15 and 29 returns to 1 % increase in discouraged worker effect across all 15 study regions on average. The second largest effect on discouraged effect is found from regional fixed effect listed in Table 4 for model (3) that ranges from 0.0751 in Chungnam to 0.0902 in Gangwon. It was expected to show that improving labor market conditions could reduce discouraged worker effects. For instance, decrease in unemployment rate of young workers and/or that of college graduate (*UEMR_1529_LG* and/or

UEMR_CG_LG) and increase in new hires standardized by job openings (*RJAS_all_LG*) are expected to improve labor market efficiency by reducing discouraged worker effect of a region. However, proposed regional-temporal fixed effect model does not support this idea; none of the unemployment variables and new hires variable is statistically significant. Expected sign for temporally lagged unemployment variables of young workforce (*UEMR_1529_LG*) and/or that of highly educated workforce (college graduate, *UEMR_CG_LG*) is positive. Though not significant, the coefficients for all unemployment variables are negative for model (3). The only significant coefficient for unemployment rate for a fixed model is from variable, *UEMR_CG_LG*, in model (2) at 5 % level. This can be interpreted as follows: improving labor market condition for highly educated workforce further increases overall discouraged workers in a region. In other words, filtering process and/or ripple effect in a regional labor market is not in place. Rather, this may contribute to increasing competition in a labor market; accordingly those with relatively lower skill level and/or education attainment tend to be pushed out for economic reason, too much competition. Other significant variables include labor force participation rate for young workers (*LFPR_1529*) and that for college graduate or higher, highly educated workforce (*LFPR_CG*), and these have a negative sign as expected. For the former, 10 % increase in labor force participation rate of population aged between 15 and 29 translates into 0.567 % decrease in the share of discouraged in a regional labor market. For the latter, 10 % increase in labor force participation rate of population with education attainment of college graduate or higher reduces 0.575 % of discouraged worker share in a region. This is intuitive and, more importantly, among all other age cohorts and groups of people with various education attainments, only these two groups, young workforce and/or highly educated workforce, can have verified impact to resolve discouraged worker effects. Their active involvement in labor market can improve overall labor market efficiency by reducing discouraged workers in a region. Migration does not play an important role on discouraged workers of a region. With pooled OLS, the coefficient of migration efficiency variable (*MIGER_all*) is positive, 0.0150, and significant at 5 % level. This reveals that a region with higher positive net migration tends to have a higher share of discouraged workers. In line with higher net migration and migration efficiency in SMA (especially with increasing population through migration in Incheon MA and Gyeonggi province), discouraged worker is a growing concern. However, there is no clear evidence between migration efficiency and discouraged worker in a regional labor market when controlling regional and temporal fixed effects. Temporal dummy variables for model (3) are all significant. Their impact on discouraged workers is bigger for 2004 through 2014, compared to that for 2003. Noticeable temporal effect is between 2009 and 2012 during which coefficients are much higher than other years. This is the period when Korean labor market had struggled facing global economic downturn.

5 Conclusion

Most conventional labor underutilization measures tend to underestimate labor underutilization by excluding discouraged workers due to economic condition from labor force. Korea is not an exception and has growing concerns on lower labor force participation of younger and/or highly educated workforces. This is partly due to the structural change in Korean economy, such that it has transformed from labor-intensive industrial structure to knowledge-based and less labor-intensive one. Domestic consumption in Korea composes only 50 % of the total GDP, while the rest is from trade with open market economy. For this reason, Korean economy is vulnerable to external shocks and domestic consumer market is closely linked. Labor force with relatively lower skill level from lower education attainment is mainly concentrated in service industries that require lower skill level. More importantly, those young workforces even with higher level of education attainment suffer from the lack of relevant types of employment opportunities with much higher barriers to enter job market. For these workforces, two options are available: (1) underemployed/unemployed or (2) withdrawal from labor force (discouraged workers). Both greatly hamper labor market efficiency in Korean economy. Moreover, the excessive concentration of various opportunities including jobs within SMA makes matters more complicated. The limited job opportunities for experienced high-skilled workers center on SMA, and this greatly discourages younger workforces in ROK even with relatively high education attainment level. The proposed models in this study specify how the share of discouraged workers in a region is determined by various factors such as demographic structures, labor market conditions, and migration pattern with regional and/or temporal fixed effects. The regional-temporal fixed effect model found to be the best in explaining how these factors determine the relative stock of discouraged workers in a region. Among the demographic structure variables in a region, the higher concentration of young population aged between 15 and 29 in a region increases the overall discouraged workers in the region, whereas all the other variables including higher education attainment level do not explain the regional stock of discouraged workers. From labor market condition variables, better labor market conditions measured by a lower unemployment rate and the higher ratio of new hires to job seekers of a region do not reduce the stock of regional discouraged workers. Only two factors describing regional labor market conditions, labor force participation rate of young cohort (aged between 15 and 29) and that of highly educated population (college graduate or higher), show direct influence on a regional stock of discouraged workers. If these groups more actively participate in labor market, the overall discouraged worker effect in a region reduces, leading to enhanced labor market efficiency. Migration pattern does not play any role on regional discouraged workers. This indicates the limited role of interregional migration on factor price equalization among regional labor markets in Korea. Regional and temporal fixed effects are found to be important. Especially, temporal effect was greater and evident from 2009 to 2012 due to global economic crisis. Regional fixed effect clearly reveals the list of regions performing better or worse than labor market of Seoul. About half of the regions perform better than Seoul including Daejeon MA and Chungnam province, the direct beneficiaries from relocation of executive branch of the Korean government into a newly planned city of Sejong. Also, regional economy in Chungnam province has grown due to the presence of strong R&D sectors and the increasing trade with China. Among the regions that perform worse than Seoul, Busan MA and Gangwon province are noticeable. Especially, Busan, the second largest city and the closed port city to Japan, has been significantly suffered from the decline of labor-intensive manufacturing activities since the mid-1980s. Gangwon province traditionally lacks sustainable industrial base and instead is known to be tourist destination with the concentration of seasonally varying service industries.

National policies to enhance overall efficiency of labor market should be designed to directly increase labor force participation of younger population and highly educated workforces. These are the two key groups of population to reduce the discouraged worker effects. Even with improving labor market conditions for these groups, these groups are not easily motivated to go back to labor force. Variables to measure improving labor market conditions do not consider the quality of jobs; instead these only focus on the number of jobs. For these groups, it is critical where to start their career paths when entering labor market. Especially, those younger populations with higher education attainment are reluctant to be underemployed and/or unemployed. Rather, they tend to stay away from labor force by quitting job search, and this is the typical case of NILF (not-in-labor force) due to economic reasons. Accordingly, policy instruments should focus on identifying growing issues of skill mismatch between overqualified labor supply and limited labor demand for these groups. From a regional perspective, policies to attract and/or locally grow highly skilled workers among young population may further aggravate regional labor market efficiency. It is strongly required for regional government to closely supervise regional industries and businesses if they provide adequate levels of salaries for younger population. Heightened barriers for these young and highly educated populations to enter labor market should be restructured. It is critical to broaden opportunities without sacrificing their returns to labor, since not just the number of jobs, but the quality of jobs matters.

Empirical models in this study utilize data from the Korean *Ministry of Employment and Labor* (MOEL) with many labor statistics, somewhat limited for various age groups and education attainment levels. Regional job openings, new hires, and job seekers statistics by age and education are relatively new, only available since 2008, and MOEL quits releasing data on job openings by age groups since 2013. With more data accumulation, the panel data regression with fixed effect as suggested in this study will be able to investigate a wider variety of factors from labor market conditions. Though controlled by regional and temporal fixed effects, other model specifications directly considering regional industrial structures and interregional migration may have higher explanatory powers on discouraged worker effects. This extension of this study can be made on how regional and temporal fluctuation of discouraged workers contributes to regional and national labor market efficiency.

References

Bal, F., and P. Nijkamp. 1997. *Exogenous and endogenous spatial growth models*, TI Discussion Paper 97-022/3. Amsterdam: Tinbergen Institute.

Barr, A., and S.E. Turner. 2013. Expanding enrollments and contracting state budgets the effect of the great recession on higher education. *The ANNALS of the American Academy of Political and Social Science* 650(1): 168–193.

Barro, R.J., and X. Sala-i-Martin. 1995. *Economic growth*. New York: McGraw-Hill.

Basker, E. 2002. *Education, job search and migration*. University of Missouri-Columbia Working Paper, (02–16).

Berry, C.R., and E.L. Glaeser. 2005. The divergence of human capital levels across cities. *Papers in Regional Science* 84(3): 407–444.

Blanchard, O.J., and P. Diamond. 1994. Ranking, unemployment duration, and wages. *The Review of Economic Studies* 61(3): 417–434.

Blundell, R., and S. Bond. 1998. Initial conditions and moment restrictions in dynamic panel data models. *Journal of Econometrics* 87(1): 115–143.

Clark, K.B., L.H. Summers, C.C. Holt, R.E. Hall, and M.N. Baily. 1979. Labor market dynamics and unemployment: A reconsideration. *Brookings Papers on Economic Activity* 1: 13–72.

Coibion, O., and Y. Gorodnichenko. 2013. *Is the phillips curve alive and well after all? Inflation expectations and the missing disinflation*. (No. w19598). National Bureau of Economic Research.

Costa, D.L., and M.E. Kahn. 2000. Power couples: Changes in the locational choice of the college educated, 1940–1990. *The Quarterly Journal of Economics* 115(4): 1287–1315.

David, H., N. Maestas, K.J. Mullen, and A. Strand. 2015. *Does delay cause decay? The effect of administrative decision time on the labor force participation and earnings of disability applicants*. (No. w20840). National Bureau of Economic Research.

Elhorst, J.P. 2000. *The mystery of regional unemployment differentials: A survey of theoretical and empirical explanations. Research report* no 00C06, SOM. Groningen: University of Groningen.

Elsby, M.W., B. Hosijn, and A. Sahin. 2013. *On the importance of the participation margin for labor market fluctuations*. Unpublished manuscript. Available at SSRN: http://nyfedeconomists.org/sahin/40483_EHS.pdf.

Faggian, A., P. McCann, and S. Sheppard. 2006. An analysis of ethnic differences in UK graduate migration behaviour. *Annals of Regional Science* 40(2): 461–471.

Faggian, A., P. McCann, and S. Sheppard. 2007. Some evidence that women are more mobile than Men: Gender differences in UK graduate migration behaviour. *Journal of Regional Science* 47 (3): 517–539.

Finegan. TA. 1978. The measurement, behavior, and classification of discouraged workers, *National commission on employment and unemployment statistics*. Washington DC: U.S. Govt. Print Office.

Flinn, C., and J.J. Heckman. 1983. The likelihood function for the multistate-multiepisode model in 'models for the analysis of labor force dynamics'. *Advances in Econometrics* 2: 225–231.

Ghayad, R., and W.T. Dickens. 2012. What can we learn by disaggregating the unemployment-vacancy relationship?. *FRB of Boston Public Policy Brief*, (12–3).

Gönül, F. 1992. New evidence on whether unemployment and out of the labor force are distinct states. *Journal of Human Resources* 27(2): 329–361.

Gordon, R.H., and A.L. Bovenberg. 1996. Why is capital so immobile internationally? Possible explanations and implications for capital income taxation. *American Economic Review* 86(5): 1057–1075.

Gottlieb, P.D., and G. Joseph. 2006. College-to-work migration of technology graduates and holders of doctorates within the United States. *Journal of Regional Science* 46(4): 627–649.

Greenwood, M. 1975. Research on internal migration in the United States: A survey. *Journal of Economic Literature* 13: 397–433.

Haugen, S.E. 2009. *Measures of labor underutilization from the current population survey*, BLS Working Paper 424. Washington, DC: US Bureau of Labor Statistics. Retrieved March 2009, from http://www.bls.gov/ore/pdf/ec090020.pdf.

Jones, S.R., and W.C. Riddell. 1999. The measurement of unemployment: An empirical approach. *Econometrica* 67(1): 147–162.

Kaldor, N. 1970. The case for regional policies. *Scottish Journal of Political Economics* 17(3): 337–348.

Kodrzycki, Y. 2001. Migration of recent college graduates: Evidence from the national longitudinal survey of youth. *New England Economic Review* 2001: 13–34.

Kroft, K., F. Lange, M.J. Notowidigdo, and L.F. Katz. 2014. *Long-term unemployment and the great recession: the role of composition, duration dependence, and non-participation.* (No. w20273). National Bureau of Economic Research.

Kubo, Y., and H. Kim. 1996. *Human capital, imported technology and economic growth: a comparative study of Korea and Japan*, Discussion paper 695, Institute of Policy and Planning Sciences, University of Tsukuba.

Kuch, P., and S. Sharur. 1978. Added- and discouraged-worker effects in Canada, 1953–1974. *Canadian Journal of Economics* 11(1): 112–120.

López-Bazo, E., T.D. Barrio, and M. Artis. 2002. The regional distribution of Spanish unemployment: A spatial analysis. *Papers in Regional Science* 81(3): 365–389.

Lucas, R.E. 1988. On the mechanics of economic development. *Journal of Monetary Economics* 22(3): 3–42.

Mathur, V. 1999. Human capital-based strategy for regional economic development. *Economic Development Quarterly Econ* 13(4): 203–221.

Nijkamp, P., and J. Poot. 1998. Spatial perspectives on new theories of economic growth. *Annals of Regional Science* 32(1): 7–37.

Okun, A.M., W. Fellner, and A. Greenspan. 1973. Upward mobility in a high-pressure economy. *Brookings Papers on Economic Activity* 4(1): 207–261.

Ondeck, C. 1978. Discouraged workers' link to jobless rate reaffirmed. *Monthly Labor Review* 101 (10): 40–42.

Romer, P. 1989. *Endogenous technological change*, No. w3210. Cambridge, MA: National Bureau of Economic Research.

Schachter J.P., R.S. Franklin, and M.J. Perry. 2003. *Migration and geographic mobility in metropolitan and non–metropolitan America: 1995–2000.* Census 2000 Special Reports, CENSR-12.

Schwartz, A. 1973. Interpreting the effect of distance on migration. *The Journal of Political Economy* 81(5): 1153–1169.

Sjaastad, L. 1962. The costs and returns of human migration. *Journal of Political Economy* 70: 80–93.

Sorrentino, C. 1995. International unemployment indicators, 1983–1993. *Monthly Labor Review* 118: 31.

Tachibanaki, T., and K. Sakurai. 1991. Labour supply and unemployment in Japan. *European Economic Review* 35(8): 1575–1587.

Van Dijk, J., H. Folmer, H.W. Herzog Jr., and A.M. Schlottmann. 1989. *Migration and labour market adjustment*. Dordrecht: Kluwer.

Waldorf, B.S. 2009. Is human capital accumulation a self-propelling process? Comparing educational attainment levels of movers and stayers. *Annals of Regional Science* 43(2): 323–344.

Part II
Industrial Agglomeration and Regional Policies

Innovation of ICT Manufacturing and Agglomeration Economies: Evolution over the Life Cycle

In Kwon Park and Gyuhwan Kim

Abstract Innovation is one of the most important driving forces of ICT industries, the leading industry of the Korean economy, and agglomeration economies are known to have positive impacts on innovation. The evolutionary agglomeration theory, however, suggests that the effects of agglomeration economies may vary depending on the life cycle of an industry. This study investigates how agglomeration economies have affected the innovation of the ICT manufacturing sector over its life cycle in the Seoul Capital Area (SCA), South Korea. A panel data set for patent applications during the period 2001–2013, which falls into the birth stage (2001–2003), the growth I (2003–2006) and II (2006–2009) stages, and the maturity stage (2009–2013), is used to model innovation in terms of various types of agglomeration economies. The results show that the types of agglomeration economies that have significant impacts on innovation vary over the life cycle of the industry: while local industrialization has a positive impact only in the birth stage, diversity has positive impacts in all the stages except the growth II stage. While large firms' leading has a positive impact only in the growth II stage, competition among small firms has a positive impact only in the birth stage. This implies that different strategies for agglomeration are needed over the life cycle in order to sustain innovation in ICT manufacturing.

Keywords ICT manufacturing • Innovation • Industry life cycle • Patents • Agglomeration economies • Evolutionary agglomeration theory

1 Introduction

It has been well acknowledged that innovation and technological growth is a driving force of the growth of an economy, since Solow (1956) introduced his seminal model on economic growth. In particular, the information and communication technology (ICT) industry as a knowledge-based industry depends heavily

I.K. Park (✉) • G. Kim
Department of Urban Administration, University of Seoul, Seoul, Korea
e-mail: ikpark@uos.ac.kr; therute1@uos.ac.kr

E. Kim, B.H.S. Kim (eds.), *Quantitative Regional Economic and Environmental Analysis for Sustainability in Korea*, New Frontiers in Regional Science: Asian Perspectives 25, DOI 10.1007/978-981-10-0300-4_4

on knowledge growth, and thus innovative activities are crucial to its growth. The intensity of knowledge production in the ICT sector is confirmed by its high number of patent applications. For instance, the number of patents applied by the ICT manufacturing sector accounts for 21.5 % of the total patents applied in Korea in the year 2014, while the number of employees of the sector accounts only for 4.1 % of the total employees in the same year.[1]

The theory of agglomeration economies suggests that agglomeration has positive impacts on innovation, and a lot of empirical studies support the theoretical expectation. There is even some evidence that the importance of agglomeration economies in knowledge production is growing (Sonn and Park 2011). The theory explains that geographical proximity among researchers in an agglomeration facilitates exchange of tacit knowledge through face-to-face communication, leading to knowledge spillovers. This is the case in particular when knowledge or technologies are too complex and tacit to be standardized. As knowledge becomes codified and can be readily transmitted for a long distance, however, the importance of geographical proximity may diminish. This implies that the impact of agglomeration on innovation may diminish as an industry becomes mature.

The evolutionary agglomeration theory indeed suggests that the effects of agglomeration vary with the life cycle of an industry. The theory maintains that externalities from geographical proximity between economic actors matter only in the early stage of the life cycle, as knowledge and technologies are tacit and non-standardized in this stage. There are a considerable number of empirical studies that support this theoretical expectation (Audretsch and Feldman 1996; Nesta and Mangematin 2002; Potter and Watts 2011).

As agglomeration economies may have different types depending on the context of a local economy, however, it may be imprudent to conclude that the effects of agglomeration indiscriminately diminish with the industry life cycle. It is known that there are many different types of agglomeration economies affecting the productivity of firms and innovation. Industrial localization and diversity have different mechanisms to affect the local economy, all constituting agglomeration economies. Competition of small firms and large firms' leading in a locality are known to be different types of agglomeration economies. These different types of agglomeration economies may have different impacts on innovation over the industry life cycle. For instance, industrial localization, i.e., geographical concentration of firms in the same sector, may have positive impacts on innovation only in the early stages, as explained by the evolutionary agglomeration theory. But industrial diversity may still have some positive impacts on innovation even in the later stages, as new ideas and creative atmosphere from industrial diversity tend to be confined to a limited distance.

To our knowledge, however, there are only few studies that separately investigate the impacts on innovation of different types of agglomeration economies over

[1] The data for patents come from the Intellectual Property Statistics for 2014 by the Korean Intellectual Property Office (KIPO 2015a), and the data for employees come from the Labor Demand Survey by the Ministry of Employment and Labor (MOEL 2015).

the industry life cycle. Kim and Park (2015) as a rare exception try this disentanglement by analyzing the effects of different types of agglomeration in two different stages of the life cycle, the growth stage versus the maturity stage of the ICT sector, using data on patents in the Seoul Capital Area (SCA), South Korea. This study extends that study by covering the whole time period from the birth to maturity stages and expands on the literature on the evolutionary agglomeration theory.

This chapter is organized as follows: Sect. 2 reviews the literature on the relationship between innovation and agglomeration economies and introduces the evolutionary agglomeration theory. Section 3 introduces the study area, the SCA, and reviews the development of the ICT manufacturing sector and the current status of innovation in the study area. Section 4 introduces the data and methods used in the analysis. Section 5 presents and discusses the results of the model estimates. Lastly, Sect. 6 concludes with a review of the findings and suggests policy implications.

2 Theory and Literature Review

2.1 Innovation and Agglomeration

It is well acknowledged that agglomeration of economic activities contributes to innovation, i.e., the production of new knowledge through knowledge spillovers. Thus, the knowledge spillovers in an agglomeration are often called "innovation externalities" (Johansson 2005). Innovation externalities are associated with the firm's development activities such as R&D efforts that lead to the dynamic process of a change in productivity. A large economy accumulates a large amount of knowledge and information which spills over from firms to firms. It helps individual firms innovate in terms of developing new products and improving their production procedures. This implies the savings of adaptive costs that are associated with learning and acquiring new technologies and developing new products and/or procedures.

Many scholars have regarded geographical proximity as the main source of external economies. When Marshall suggested external economies of agglomeration, he assumed the concept of "industrial districts" where small firms are locally concentrated and locally make a decision and produce (Markusen 1996). When Hoover classified external economies into localization economies and urbanization economies, he also presumed that firms of the same industry or diverse industries locate within a locality. Urban and regional economists such as Hirschman and Isard also focused on agglomeration economies (Polenske 2001). Krugman and Fujita's seminal works on new economic geography also paid attention to it (Fujita and Krugman 2004).

Knowledge spillovers are generated a by-product of various interactions among economic actors in a dense environment, as geographical concentration of economic activities implies physical proximity between actors in the local market. Two mechanisms through which the physical proximity leads to knowledge spillovers are suggested: First, large local demands generate information about diverse demands and this in turn stimulates innovation in local firms. In other words, diverse demand in the local market itself is a good source of information. Second, proximity allows local firms to observe their competitors and imitate best practice, even when there is no explicit cooperation among competitors. In either case, proximity plays a key role because tacit knowledge and complex information are created and transmitted mainly by face-to-face contact (Jaffe 1989; Jaffe et al. 1993; Feldman 1994; Mansfield 1995; Anselin et al. 1997, 2000; Audretsch 1998; Lim 2004; Johansson 2005; Storper and Venables 2004; McCann 2007; Henderson 2007).

Given that geographical proximity of economic actors can lead to knowledge spillovers, it matters how the local market is composed. The effect on innovation of agglomeration may vary depending on the composition of local actors. One of the long controversies is about which of specialization and diversity is more favorable to innovation (Feldman and Audretsch 1999). The Marshall-Arrow-Romer (MAR) theory suggests that industrial localization, which refers to concentration of firms of the same sector within a region, is more favorable to knowledge spillovers (Glaeser et al. 1992). Porter (1990) also emphasizes the localization economies that clusters of the same industry benefit from externalities such as knowledge spillovers. On the other hand, Jacobs (1969) advocates urbanization economies arguing that industrial diversity pays more than localization. New ideas can arguably be created through the exchange of diverse knowledge across industries. Boschma et al. (2015) are also on Jacobs' side showing that technological relatedness leads to industrial diversification, which contributes to creating new ideas.

Another controversy is about which of competition and monopoly is more favorable to innovation. Regarding this question, the MAR theory advocates monopoly arguing that too competitive a market condition makes firms differentiate their prices while imitating their rivals' production rather than creating new ideas. On the contrary, Porter (1990) advocates competition among small firms as fierce competition stimulates their innovation.

Many empirical studies support both industrial localization and diversity and both monopoly and competition. The bottom line is that the types of agglomeration economies that have significant impacts on innovation vary depending on the region, industry, and economic context.

2.2 Evolution of Agglomeration Economies

The life cycle of an industry is one of the economic contexts that determine the impacts on innovation of agglomeration, as the innovative activity tends to be shaped by the stage of the life cycle (Klepper 1996; 1997). Evolution agglomeration theory suggests that the life cycle determines whether or not agglomeration leads to increasing returns. As an industry moves along the stages of its life cycle – the embryonic, growth, maturity, and decline stages according to Vernon (1966)– agglomeration generates increasing returns only in the first two stages, followed by constant returns and diminishing returns in the following stages (Potter and Watts 2011).

The positive externalities of agglomeration diminish with time, as knowledge becomes standardized and codified, and thus the cost of transmission for a long way becomes smaller (Audretsch 1998). The externalities can become even negative with time, as firms are locked in outdated technologies in an agglomeration. As reviewed by Kim and Park (2015), there are a considerable number of empirical studies supporting such an expectation of evolutionary agglomeration theory (Audretsch and Feldman 1996; Nesta and Mangematin 2002; Potter and Watts 2011).

As mentioned earlier, the concrete shapes of agglomeration economies are varied as agglomeration can have various compositions of economic actors. This complicates the problem and makes it inappropriate to conclude that all types of agglomeration economies indiscriminately diminish with time. In order to look into the detailed story, we should disentangle the effects of agglomeration by its type over the life cycle of an industry. Industrial diversity may have positive impacts on innovation even when industrial localization has no longer significant impacts at all, as diverse knowledge sources help produce new ideas. Competition among small firms is more favorable to innovation than monopoly or some large firms' leading, when the industry is in its early stage of the life cycle. But the opposite may be true when many small firms shut down due to harsh competition, and their researchers are absorbed by some large firms. Various combinations of the effects of agglomeration are possible. To figure out the detailed story, one should separate the effects of agglomeration economies by type over the life cycle of an industry. Unfortunately, however, only a few studies try the disentanglement, in particular in relation to the effects on innovation.[2]

Kim and Park (2015) as a rare exception try to disentangle the effects on knowledge production of agglomeration economies by its type and the industry life cycle, using the case of ICT industries in the Seoul Capital Area, South Korea. Due to the limitation of data, however, they just investigate the stages of later growth and maturity. This study is an extension to Kim and Park (2015) in the time dimension so that it covers the whole stages that industry has gone through, from the birth stage to the maturity stage. Also, this study measures the impacts on innovativeness, as measured by the number of patents per ICT worker, rather than on the amount of knowledge production.

---

[2] For multiple dimensions of agglomeration economies in general, see Park and von Rabenau (2011). They disentangle the externalities of agglomeration by agent, source, and spatial dependence.

3 Development and Innovation of ICT Manufacturing in the SCA

3.1 ICT Manufacturing in the SCA

The ICT manufacturing sector[3] occupies the largest fraction of R&D expenditures by business enterprises. As of 2013, businesses in the ICT manufacturing sector in Korea spent about 25 billion 2005 USD, which is equivalent to half of the total R&D expenditures of all businesses (OECE.stat 2015).[4] Since one can say that innovation is generated where R&D spending is high (Griliches 1990), the ICT manufacturing sector has many innovative activities and is a good industry to study innovation.

The Seoul Capital Area (SCA) in South Korea is a good area to study innovation in ICT manufacturing and agglomeration economies, as the region is populated by a lot of firms and employees in the sector, making one of the largest ICT clusters in the world. Also, the region is home to about 25 million people and produces about 636 billion USD[5] of products as of 2013, which is 48.7 % of the GDP of the country (KOSTAT 2015a), and thus it plays a role of the economic center of the country. For details, see Kim and Park (2015).

The study covers the period 2001–2013, when the ICT manufacturing sector went through the life cycle from the birth stage to the maturity stage. Although the birth stage of the ICT manufacturing began in the mid-1990s, we just investigate only this period due to a limitation of patent data. Kim and Park (2015) divide the life cycle of the Korean ICT manufacturing sector into three stages – birth stage (1990–2002), growth stage (2002–2009), and maturity stage (2009–present) – as the sector has not entered the decline stage yet. The production trend of the sector supports the division scheme as shown in Fig. 1. The production, measured by production index compared to the total production in 2010, grew slowly in the birth stage and then rapidly before the global financial crisis in 2008. After that, however, the growth was slowed down. Between the birth stage and the growth stage, the country co-held with Japan a world-renowned event, the FIFA World Cup Games in 2002, which was an occasion to boost the ICT sector. After the event, the industry began to grow rapidly and a lot of start-ups entered business and the number of firms and employees increased a lot. Also, ICT clusters were formed in the SCA in

[3] The ICT manufacturing sector in this study refers to the industries assigned to the code C26 (manufacture of electronic components, radio, television, and communication equipment and apparatuses) by the Korean Standard Industrial Classification (KSIC), Rev. 9, which includes the manufacture of semiconductor (C261), electronic components (C262), computers and peripheral equipment (C263), telecommunication and broadcasting apparatuses (C264), electronic video and audio equipment (C265), and magnetic and optical medium (C266).

[4] ICT manufacturing here refers to the industries assigned to the International Standard Industrial Classification (ISIC), Rev. 4 codes 261, 262, 263, and 268.

[5] The exchange rate 1 USD = 1095.6 KRW is applied.

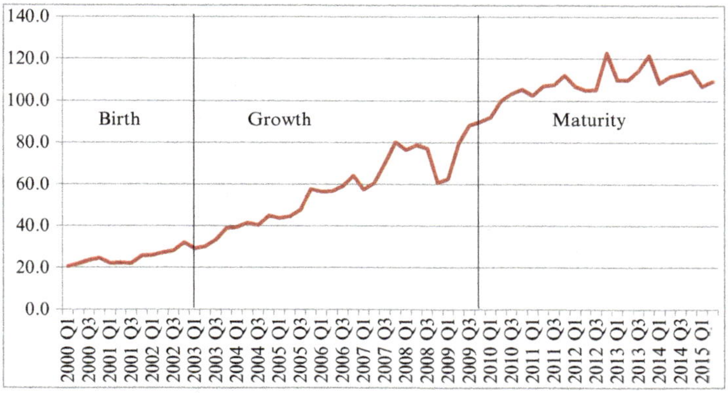

Fig. 1 Quarterly data on production in ICT manufacturing in Korea (Data source: KOSTAT 2015b)

the growth stage. While going through the global financial crisis, the sector lost many small firms that were merged into large conglomerates. In other words, there is a kind of restructuring period during the later 2000s after the rapidly growing stage. After the occasion, the sector's growth was slowed down and entered the maturity stage. See Kim and Park (2015) for further information.

Unlike Kim and Park (2015), however, we divide the years 2003 and 2009 as the turning points of the life cycle of the ICT manufacturing sector. As there are usually time lags between economic shocks and their effects, the effects on innovation of the FIFA World Cup Games in 2002 and the global financial crisis in 2008 were realized in 2003 and 2009, respectively. This lagged pattern is also found in the development of innovation in ICT manufacturing; see Sect. 3.2 for details.

3.2 *Innovation in ICT Manufacturing in the SCA*

The growth of technology has been known as the growth engine since Solow (1956) introduced the seminal model of economic growth (Acs and Varga 2002). Ultimately, innovation is essential in economic growth, as the growth of technology is generated by innovation. This is also the case for the ICT manufacturing sector in the SCA, South Korea. Innovation has been in line with the growth of the industry, with a high level of innovation during the period of rapid growth.

This study measures the level of innovation using the number of patents, following Griliches (1990) and many subsequent studies (Crépon et al. 1998; Porter and Stern 1999; Mairesse and Mohnen 2002; Acs et al. 2002; Mairesse et al. 2005). Griliches finds that the number of patents is highly correlated to R&D expenditures both in the cross-sectional dimension across firms and in the within-firm time-series dimension. Given the readiness of plentiful patent data, he concludes that patents

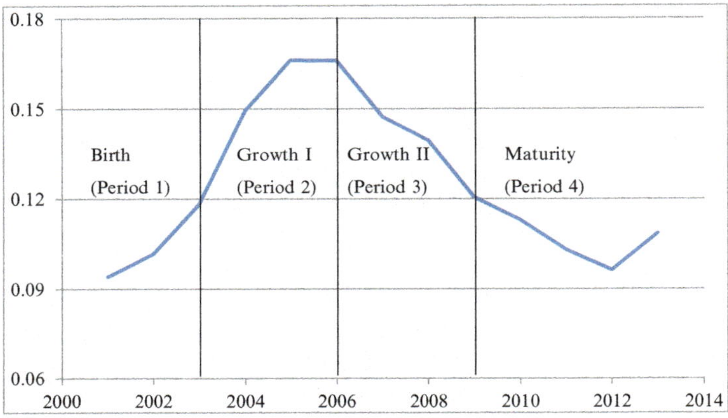

Fig. 2 Innovation in ICT manufacturing in the SCA (Data: Courtesy of the Korean intellectual property office)

are a good resource for analyzing technical change. In particular, we adopt the number of patents per worker, which is used by Crépon et al. (1998) as one of two alternative measures of innovation output.[6] Since the number of patents is normalized by the number of workers in the industry, the measure represents the degree of innovativeness.[7]

Figure 2 shows changes in innovation level in the ICT manufacturing sector, measured by the number of patent applications per worker in the SCA. The innovation slowly increased until the year 2003, and this period corresponds to the birth stage of the ICT manufacturing in Korea. The period 2003–2009, corresponding to the growth stage, saw a rapid growth of innovation. However, there is a distinct pattern discovered within this period. In the first half of the period (2003–2006), the innovation in the sector grew, while in the second half (2006–2009), the innovation level declined but still stayed at levels higher than during the other two stages. Given this distinction, we divide the growth stage into two substages: growth I and growth II. Overlapping the innovation trend with the production trend, one can say that the growth II stage is a kind of transition period between the rapidly growing stage and the maturity stage. The period since the year 2009 has seen a further decline in the innovation level with a pickup in 2013.

[6] The other measure suggested by Crépon et al. (1998) is the share of innovative sales.

[7] If it is not defined otherwise, innovation refers to the degree of innovativeness in this study.

4 Data and Methods

4.1 Data

We examine innovation in terms of patents in the ICT manufacturing sector in the Seoul Capital Area (SCA), South Korea. For the research, we use a panel data set consisting of 60 municipalities and 13 years (2001–2013). The spatial unit of analysis is the municipality, Si-Gun-Gu, for which data are the most available in the country, and we chose 60 Si-Gun-Gu's omitting six due to the missing data. [8]

As mentioned earlier, we measured innovation in terms of the number of patents applied per worker for a year. The data for patent applications come from the Statistic for Applied Patents by Industry (SAPI) that was provided by the Korean Intellectual Property Office (KIPO). The place of innovation is identified by the applicant's address,[9] and the degree of innovation for a municipality is calculated by dividing the number of all the patent applications by the number of employees in ICT manufacturing for a year. In this data set, however, patents are classified according to the International Patent Classification (IPC) system, which is different from the Korean Standard Industrial Classification (KSIC) system. In order to identify the patents in ICT manufacturing, we matched up IPC codes with KSIC codes, according to the matching table provided by KIPO (2015b). Therefore, the number of patent applications for a region is figured out by summing up all the numbers for the IPC codes in Table 1. Also, as there was a revision from Rev. 8 to Rev. 9 for the KSIC code during the period of study, we matched up the two different systems based on the matching table provided by KOSTAT (2008a), as shown in Table 1.

The data for the other variables, such as the establishments, population, and employment, were collected from the official statistics warehouse, the Korean Statistical Information Service (KOSIS). These data are retrieved for individual municipalities and joined with the patent data to form the complete data set.

4.2 Model and Variables

We introduce a regression model to determine the effect on innovation of agglomeration economies over the industry life cycle. The model regresses the number of patent applications per worker (*innov*) in the form of natural logarithm on a set of explanatory variables that represent four types of agglomeration economies for four periods, Period 1 (birth, 2001–2003), Period 2 (growth I, 2003–2006), Period 3 (growth II, 2006–2009), and Period 4 (maturity, 2009–2013).

[8] The omitted six municipalities are Gapyeong-gun, Yeonsu-gu, Yangpyeong-gun, Yeoncheon-gun, Ganghwa-gun, and Ongjin-gun.

[9] For an application, two addresses are reported: inventor's and applicant's. As the addresses for inventors include only those of the registered inventors such as institutions, we use the more comprehensive applicants' addresses.

Table 1 Matching of industry classifications between KSIC and SAPI

Classification of ICT industries by KSIC			Classification of ICT patents and utility models by SAPI
Rev. 8	Rev. 9	Name	IPC Code
D321	C261	Semiconductor manufacturing	G11C, H01L
	C262	Electronic component manufacturing	B05D, B81B, B81C, B82B, B82Y, C30B, G02F1, G12B, H01C, H01F, H01G, H01J, H05K, H03K, H03L
D300	C263	Computer and peripheral device manufacturing	B41J2, B41J27, B41J29, B41J31, B41J32, B41J33, B41J35, G06J, G06F1, G06F3, G06K
D322	C264	Communication and broad-cast equipment manufacturing	G03H, H01P, H01Q, H01S, H03B, H03C, H03D, H03F, H03G, H03H, H03M, H04B, H04J, H04K, H04L, H04M, H04Q, H04S, H04W, H99Z
D323	C265	Image and sound equipment manufacturing	G03B31, G10L, H03J, H04H, H04N, H04R
D333	C266	Magnetic and optical medium manufacturing	G11B

Source: KOSTAT (2008) and KIPO (2015b)
Note: Since KIPO opens data only for patents with IPC codes of up to four digits, the subgroups under a four-digit IPC code were ignored, and their data were collected with the four-digit code. For instance, the number of patents with IPC codes B41J2, B41J27, B41J29, B41J31, B41J32, B41J33, and B41J35 is figured out by the number of patents with IPC code B41J. The same applies to G06F1, G06F3, and G03B31

The four types of agglomeration economies in the right hand side of the model are industrial localization, diversification, competition among small firms, and the large firms' leading. Industrial localization is measured by a cluster index (*clu*), and the degree of industrial diversity is measured by the inverse of the Herfindahl-Hirschman index for all industries (*div*), for the manufacturing industry only (*div_mfg*), and for ICT manufacturing only (*div_ict*). Competition among small firms is measured by a region's number of ICT manufacturing firms per worker relative to the country's counterpart (*compe*). The large firms' leading is measured by the proportion of ICT workers in the region who are employed by large firms with 300 employees or more (*large*).

Other factors are controlled for in the model including the human resource factor. The employment rate (*emp_rat*) is included to consider the economic vitality of the region. Human resource factors such as ICT workers (*ln_ictemp*), researchers (*res_rat*), and population (*ln_pop*) are included as well. The definitions and measurements of the variables are presented in Table 2.

A panel regression model is introduced, as the data consist of cross sections for different time periods. The model is written as

$$y_{it} = \alpha + \beta x_{it} + u_i + \varepsilon_{it}$$

where y_{it} is the dependent variable, x_{it} is the vector of explanatory variables that vary over time and region, u_i represents the unobserved time-invariant regional

Table 2 Definitions and measurements of variables

Category	Factors	Definition and measurement of variable
Dependent variable	Innovation	Natural logarithm of patent applications per worker for a year in the ICT manufacturing sector (*innov*)
Types of agglomeration economies	ICT manufacturing localization	$clu = \frac{\frac{E_{i,ICT}}{\sum_i E_{i,ICT}}}{\frac{P_i}{\sum_i P_i}} \times \frac{\frac{F_{i,ICT}}{\sum_i F_{i,ICT}}}{\frac{a_i}{\sum_i a_i}}$ $E_{i,ICT}$: number of *ICT* employees in region *i* $F_{i,ICT}$: number of *ICT* companies in region *i* P_i: number of population in region *i* a_i: area of region *i* (sum of residential, commercial, and manufacturing areas)
	Diversity (all industries)	$div = \frac{1}{\sum_j \left(\frac{E_{ij}}{E_i}\right)^2}$ E_{ij}: number of employees of industry *j* in region *i* E_i: number of all the employees in region *i*
	Diversity (manufacturing)	$div\_mfg = \frac{1}{\sum_{j \in manufacturing} \left(\frac{E_{ij}}{E_i}\right)^2}$
	Diversity (ICT manufacturing)	$div\_ict = \frac{1}{\sum_{j \in ICT} \left(\frac{E_{ij}}{E_i}\right)^2}$
	Competition	$compe = \frac{F_{i,ICT}/E_{i,ICT}}{F_{ICT}/E_{ICT}}$ F_{ICT}: number of *ICT* companies in the country E_{ICT}: number of *ICT* employees in the country
	Large firms' leading	$large = \frac{E_{i,ICT(emp \geq 300)}}{E_{i,ICT}}$ $E_{i,ICT(emp \geq 300)}$: number of employees of *ICT* companies with 300 employees or more in region *i*
Control factors	Employment rate	$emp\_rat = \frac{E_i}{P_i}$
	ICT manufacturing employees	Natural logarithm of the number of ICT employees (*ln_ictemp*)
	Researchers	$res\_rat = \frac{E_{iR}}{E_i}$ E_{iR}: number of R&D employees in region *i*
	Population	Natural logarithm of the number of total population (*ln_pop*)

effect, and ε_{it} is the normal error term. The fixed effect model is estimated if u_i is fixed, while the random effect model is estimated if u_i is a random variable. The Hausman test based on our data set for the four periods rejects the null hypothesis that u_i is uncorrelated to the included explanatory variables. But the test does not reject the null for the model for the whole period. Thus, we adopt the fixed effect model for the four subdivided periods, while we adopt the random effect for the whole period. Even after the fixed model is estimated, auto-correlation and heteroscedasticity of the error term are found for Periods 1, 2, and 4. In order to deal with these issues, we used the feasible generalized least squares (FGLS) method for the three periods, while we used the ordinary least squares (OLS) method for Period 3.

5 Empirical Findings: Agglomeration Economies and Innovation in the SCA

5.1 Results of Estimation

We estimated the model using the data for 60 municipalities in the SCA without missing data. The results are shown in Table 3 with the relevant coefficient estimates and associated standard errors for the four periods. One can easily identify different patterns over the four periods. Look into the details for each period.

First, consider the results for Period 1 (2001–2003). In this period, ICT localization (*clu*) has a positive impact on innovation, though significance is weak.[10] Competition among small firms (*compe*) also has significantly positive impacts. Diversity in all industries (*div*) has significantly positive impacts, while diversity within ICT manufacturing (*div_ict*) and large firms' leading (*large*) have negative impacts. Employment rate (*emp_rat*) and total population (*ln_pop*) have significantly positive impacts. Note that the number of ICT manufacturing workers (*ln_ictemp*) has a negative impact on innovation. This implies the diminishing marginal productivity of labor in knowledge production in the local ICT manufacturing sector. In other words, the marginal growth in knowledge production decreases with the number of workers in the sector.

Second, consider the results for Period 2 (2003–2006). First, it is notable that ICT localization (*clu*) does not have an impact on innovation any longer. Diversity in all industries (*div*) still has significantly positive impacts, while diversity in manufacturing (*div_mfg*) has a negative impact. In this period, the proportion of researchers (*res_rat*) has a positive impact on innovation. Employment rate (*emp_rat*), total population (*ln_pop*), and the number of ICT manufacturing workers (*ln_ictemp*) have similar impacts as in Period 1.

Third, consider the results for Period 3 (2006–2009). Overall, the significance levels for the four types of agglomeration economies have decreased compared to the model for Period 2. ICT localization (*clu*), diversities of all the industry levels (*div*, *div_mfg*, *div_ict*), and competition among small firms have no significant impact on innovation. Only large firms' leading (*large*) has significantly a positive impact. It turns out that the proportion of researchers (*res_rat*) has a negative impact in this period, and the total population (*ln_pop*) has no significant impact. The other variables have similar effects as in the previous periods. Overall,

[10] The level of significance may go up if we expand the time range back to the 1990s, as the earlier period within the birth stage may see higher effects of localization than the later.

Table 3 Regression results

Variable	Period 1 (2001–2003) FGLS	Period 2 (2003–2006) FGLS	Period 3 (2006–2009) Fixed	Period 4 (2009–2013) FGLS	Whole (2001–2013) Random
clu	.0290019† (.0165471)	−.0104097 (.0149041)	−.0205197 (.0247753)	.0165348 (.0106986)	.0237297 (.0163037)
div	.0315532** (.0097325)	.0242005*** (.0045112)	.0172604 (.0170243)	.0145331* (.0060361)	.02573*** (.004528)
div_mfg	−.0172264 (.0108338)	-.0495433*** (.01013)	.0271684 (.0256966)	−.0310249** (.0110449)	−.017899 (.0122793)
div_ict	−.1257718** (.0452112)	.0429059 (.0539628)	−.0134958 (.0800457)	.0204912 (.0492581)	−.0257664 (.0445092)
compe	.0949237** (.0297651)	.0306121 (.0295445)	.0603331 (.0419912)	−.0028755 (.022884)	.0791893** (.0265962)
large	−.7704083** (.2270797)	.142226 (.2350028)	.8077345** (.2885072)	.2131288 (.1999779)	.1823512 (.2632818)
emp_rat	1.654712*** (.2212961)	1.991938*** (.2826351)	3.892624* (1.502291)	1.792244*** (.1663987)	1.573693*** (.2667345)
ln_ictemp	−.6238481*** (.0755388)	−.6550448*** (.0603739)	−.9834177*** (.1803775)	−.8575201*** (.0535428)	−.7455895*** (.1094631)
res_rat	2.744417 (5.872994)	12.73016* (5.796531)	−15.99064*** (5.022202)	10.11293*** (2.515052)	1.506259 (5.134908)
ln_pop	1.236673*** (.1365056)	1.154992*** (.1312259)	.7269286 (.7201809)	1.282444*** (.1055283)	1.068113*** (.2834282)
Constant	−15.45345*** (1.706249)	−13.94178*** (1.503495)	−7.391402 (9.299599)	−13.61437*** (1.335536)	−12.36216*** (3.375489)
Observations	180	240	240	300	780
Groups	60	60	60	60	60
Time periods	3	4	4	5	13
Wald chi-square (F-value)	511.87	526.66	23.02 (F-value)	946.60	528.51
p > chi-square (p > F)	0.0000	0.0000	0.0000	0.0000	0.0000

***$p < 0.001$, **$p < 0.01$, *$p < 0.05$, and †$p < 0.1$
Note: Standard errors
(robust standard errors in the case of Period 3) are in parentheses. The F-value rather than Wald chi-square is reported for Period 3

however, this period saw an extraordinary pattern in the relationship between innovation and the explanatory factors. This can be explained by the peculiarity of this period. As mentioned in Sect. 3.1, this period underwent a restructuring where many small firms shut down and many employees in the industry lost their jobs. This peculiarity may lead to such an abnormal pattern in innovation as well.

Fourth, consider the results for Period 4 (2009–2013). In this period, of the four types of agglomeration economies, only diversity in all industries (*div*) has a positive impact on innovation, while diversity in manufacturing (*div_mfg*) has a negative impact. Localization (*clu*), competition (*compe*), and large firms' leading (*large*) are not significant anymore. The other variables have similar impacts as in Period 2.

Lastly, consider the results for the whole period (2001–2013). In this model, diversity in all industries (*div*) and competition among small firms (*compe*) have positive impacts, while the other types of agglomeration economies have no significant impact. The other control variables have similar impacts as in Period 1.

5.2 Agglomeration Economies over the Life Cycle

Putting together the results for the four stages, we can find some interesting patterns of the effects of four different types of agglomeration economies, as shown in Table 4. Overall, the number of agglomeration economies types that have significantly positive impacts on innovation decreases with time, as suggested by evolutionary agglomeration theory. While local industrialization has a positive impact on the innovation of ICT manufacturing only in the birth stage, diversity has positive impacts in all the stages except the growth II stage. While large firms' leading has a positive impact only in the growth II stage, competition among small firms has a positive impact in the birth stage. Diversity in manufacturing or in ICT manufacturing has often negative impacts on innovation. The following paragraphs discuss the meanings of these results.

Industrial localization has a positive impact on innovation only in the birth stage of the ICT manufacturing sector. This implies that geographical proximity among researchers in the industry does not matter as the industry becomes mature. This is exactly in line with the evolutionary agglomeration theory. As knowledge in the industry becomes standardized and codified, it can transmit for a long distance diminishing the importance of geographical proximity.

Diversity has significant impacts on innovation in all the stages except the growth II stage, implying that diversity is a good source of innovation for ICT manufacturing. Diversity in ICT manufacturing, however, has a negative impact in the birth stage, and diversity in manufacturing has a negative impact in the later stages. In other words, diversified industrial composition in all industries stimulates the production of new knowledge in ICT manufacturing, suggesting that Jacob's

Table 4 Impacts of agglomeration economies over the life cycle

		Birth (Period 1)	Growth I (Period 2)	Growth II (Period 3)	Maturity (Period 4)
Localization		Positive	–	–	–
Diversity	All industries	Positive	Positive	–	Positive
	Manufacturing	–	Negative	–	Negative
	ICT mfg.	Negative	–	–	–
Competition		Positive	–	–	–
Large firms' leading		Negative	–	Positive	–

theory works well in the case of innovation in ICT manufacturing. However, the growth II stage, as a kind of transition period, saw an abnormal pattern that diversity has no impact on innovation. Interestingly, diversity in related industries such as manufacturing and ICT manufacturing does not help at all. It is notable that these results contradict Boschma et al.'s (2015) argument that relatedness helps technological changes.

Competition among small firms has a significant impact on innovation only in the birth stage. This suggests that small firms benefit more from knowledge spillovers from the local market (Brown and Rigby 2013) and play a key role in innovation in ICT manufacturing in its early stage. This is in line with Porter's theory that supports the clustering of small firms rather than a large firm's monopoly.

Large firms' leading has a significant impact only during the restructuring period (growth II) when a lot of small firms shut down around the global financial crisis. This is due to the fact that the researchers of closed small firms are absorbed by large firms that can usually survive economic hardship.

6 Conclusion

This study analyzes the effects on innovation of agglomeration economies over the life cycle of the ICT manufacturing sector: birth (2001–2003), growth I (2003–2006), growth II (2006–2009), and maturity (2009–2013). The results show that the effects vary with the life cycle and that overall the importance of agglomeration economies diminishes as the ICT manufacturing sector becomes mature. These results are in line with the literature on evolutionary agglomeration theory. But the study expands on the literature by finding more complex effects that different types of agglomeration economies have on innovation over the industry life cycle.

Comprehensively reviewing the results for the four stages, we can find some interesting patterns of the effects on innovation of agglomeration economies. First, industrial localization has a significant impact on innovation only in the birth stage, though the significance level is not so high. This implies that geographical proximity among researchers in the same industry does not matter as the industry becomes mature. Second, diversity in all industries has significantly positive impacts on innovation in all the stages except the growth II stage, while diversity in manufacturing or in ICT manufacturing has negative impacts: diversity within ICT manufacturing when the industry just begins, while diversity in manufacturing when the industry is rapidly growing or is mature. This suggests that the diversified composition of industries not limited to manufacturing or ICT manufacturing is very important for innovation in ICT manufacturing. Third, competition among small firms has significant impacts on innovation only in the birth stage just like industrial localization. Combined with the first finding, this implies that small firms play a key role in the innovation of ICT manufacturing in the birth stage. Fourth, large firms' leading has a significant impact only during a transition period (growth II) when a lot of small firms shut down and their researchers are absorbed by large firms.

Based on these empirical findings, we suggest some policies for promoting innovation in ICT manufacturing. First, the clustering of firms in ICT manufacturing should be encouraged mainly in the early stage when it has significant impacts on innovation. Second, it will be good to protect and promote small firms in particular in the birth stage, as they play a key role in the innovation of ICT manufacturing in the early stage of the life cycle. Third, it is always important for the innovation of ICT manufacturing to maintain and promote industrial diversity all the time. Once the industry enters the maturity stage, the only form of agglomeration economies that matters is diversity in all industry. Fourth, we also need to acknowledge that large firms play a role during the transition when the industry restructures itself to adjust to a new normal state.

Despite such new findings on innovation in the ICT manufacturing sector, a caveat is in order: one should not conclude that the above findings can be generalized to other industries. In order to confirm the possibility, we need to investigate the cases of other industries. That is reserved for subsequent studies.

Appendix 1: Descriptive Statistics

Variable		Mean	Std. dev.	Min.	Max.	Time periods	Number of obs.
innov	Period 1	−3.017629	1.513923	−5.639145	2.140066	3	180
	Period 2	−2.744071	1.537157	−5.669881	1.676803	4	240
	Period 3	−2.18766	1.680219	−5.512291	2.872134	4	240
	Period 4	−1.906382	1.857345	−6.529544	3.351682	5	300
	Whole	−2.389176	1.742652	−6.529544	3.351682	13	780
clu	Period 1	1.672744	3.234111	.0000933	21.42845	3	180
	Period 2	1.46516	3.051903	.0000933	23.72123	4	240
	Period 3	1.473138	3.324267	.0002175	25.09793	4	240
	Period 4	1.412657	3.212534	.0001718	23.02446	5	300
	Whole	1.493706	3.191456	.0000933	25.09793	13	780
div	Period 1	33.04492	7.313514	19.15775	47.08056	3	180
	Period 2	34.50096	8.137966	16.69818	56.76448	4	240
	Period 3	39.81731	8.535121	17.6746	57.33545	4	240
	Period 4	40.06973	8.416427	18.48906	59.19137	5	300
	Whole	37.25466	8.711706	16.69818	59.19137	13	780
div_mfg	Period 1	13.95222	6.362514	3.115928	27.11503	3	180
	Period 2	13.00141	5.893449	3.013242	24.4449	4	240
	Period 3	12.33691	5.896666	2.765624	26.09841	4	240
	Period 4	11.60668	5.823097	2.727195	27.13635	5	300
	Whole	12.56633	6.044792	2.727195	27.13635	13	780

(continued)

Variable		Mean	Std. dev.	Min.	Max.	Time periods	Number of obs.
div_ict	Period 1	2.779748	.7884301	1	4.680952	3	180
	Period 2	2.617045	.7718202	1	4.717073	4	240
	Period 3	2.695144	.829697	1	5.30042	4	240
	Period 4	2.662623	.8617387	1	5.30042	5	300
	Whole	2.684715	.8224017	1	5.30042	13	780
compe	Period 1	2.431592	2.497128	.1219435	17.69739	3	180
	Period 2	2.709811	2.337954	.0978488	15.89858	4	240
	Period 3	3.349214	2.949074	.0839257	15.89858	4	240
	Period 4	3.923562	3.34462	.0907789	17.99758	5	300
	Whole	3.215274	2.960445	.0839257	17.99758	13	780
large	Period 1	.2021439	.2691816	0	.8920141	3	180
	Period 2	.1915671	.2825242	0	.9456964	4	240
	Period 3	.1604602	.2770371	0	.926899	4	240
	Period 4	.1486172	.2710214	0	.9233746	5	300
	Whole	.1715199	.2745515	0	.9456964	13	780
emp_rat	Period 1	.3693149	.3609731	.1398799	2.780034	3	180
	Period 2	.377474	.3661295	.1502432	2.780034	4	240
	Period 3	.3946322	.3734753	.1502432	2.843038	4	240
	Period 4	.424876	.3850713	.1583888	2.902833	5	300
	Whole	.3967908	.3736912	.1398799	2.902833	13	780
ln_ictemp	Period 1	7.271445	1.718777	.6931472	10.32246	3	180
	Period 2	7.218343	1.71977	2.079442	10.41286	4	240
	Period 3	6.890221	1.871622	2.302585	10.38659	4	240
	Period 4	6.703199	2.006158	2.302585	10.39336	5	300
	Whole	6.980718	1.866508	.6931472	10.41286	13	780
res_rat	Period 1	.0023691	.0051609	0	.0393788	3	180
	Period 2	.0031187	.0070078	0	.041866	4	240
	Period 3	.0048024	.0101581	0	.0626197	4	240
	Period 4	.0058194	.0123113	0	.087643	5	300
	Whole	.0042705	.0096597	0	.087643	13	780
ln_pop	Period 1	12.64271	.6366047	11.16087	13.84792	3	180
	Period 2	12.6699	.6375497	11.0101	13.88215	4	240
	Period 3	12.70115	.6364824	11.0101	13.88611	4	240
	Period 4	12.73901	.6271121	11.16447	13.95367	5	300
	Whole	12.69629	.633718	11.0101	13.95367	13	780

References

Acs, Z.J., and A. Varga. 2002. Geography, endogenous growth, and innovation. *International Regional Science Review* 25(1): 132–148.

Acs, Z.J., L. Anselin, and A. Varga. 2002. Patents and innovation counts as measures of regional production of new knowledge. *Research Policy* 31(7): 1069–1085.

Anselin, L., A. Varga, and Z. Acs. 1997. Local geographic spillovers between university research and high technology innovations. *Journal of Urban Economics* 42(3): 422–448.

Anselin, L., A. Varga, and Z.J. Acs. 2000. Geographic and sectoral characteristics of academic knowledge externalities. *Papers in Regional Science* 79(4): 435–443.

Audretsch, B. 1998. Agglomeration and the location of innovative activity. *Oxford Review of Economic Policy* 14(2): 18–29.

Audretsch, D.B., and M.P. Feldman. 1996. Innovative clusters and the industry life cycle. *Review of Industrial Organization* 11(2): 253–273.

Boschma, R., P.A. Balland, and D.F. Kogler. 2015. Relatedness and technological change in cities: The rise and fall of technological knowledge in U.S. metropolitan areas from 1981 to 2010. *Industrial and Corporate Change* 24(1): 223–250.

Brown, W.M., & D.L. Rigby. 2013. *Urban productivity: Who benefits from agglomeration economies?* Statistics Canada, Economic Analysis Division.

Crépon, B., E. Duguet, and J. Mairesse. 1998. Research and development, innovation and productivity: An econometric analysis at the firm level. *Economics of Innovation and New Technology* 7(2): 115–158.

Feldman, M.P. 1994. *The geography of innovation*, vol. 2. Dordrecht: Springer Science & Business Media.

Feldman, M.P., and D.B. Audretsch. 1999. Innovation in cities: Science-based diversity, specialization and localized competition. *European Economic Review* 43: 409–429.

Fujita, M., and P. Krugman. 2004. The new economic geography: Past, present, and the future. *Papers in Regional Science* 83: 139–164.

Glaeser, E.L., H.D. Kallal, J.A. Scheinkman, and A. Shleifer. 1992. Growth in cities. *Journal of Political Economy* 100(6): 1126–1152.

Griliches, Z. 1990. Patent statistics as economic indicators: a survey (No. w3301). National Bureau of Economic Research.

Henderson, J.V. 2007. Understanding knowledge spillovers. *Regional Science and Urban Economics* 37(4): 497–508.

Jacobs, J. 1969. *The economy of cities*. London: Vintage.

Jaffe, A.B. 1989. Real effects of academic research. *The American Economic Review* 75(5): 957–970.

Jaffe, A.B., M. Trajtenberg, and R. Henderson. 1993. Geographic localization of knowledge spillovers as evidenced by patent citations. *The Quartely Journal of Economics* 108(3): 577–598.

Johansson, B. 2005. Parsing the menagerie of agglomeration and network externalities. In *Industrial clusters and inter-firm networks*, ed. C. Karlsson, B. Johansson and R.R. Stough, 107–147. Cheltenham: Edward Elgar.

Kim, G.H., and I.K. Park. 2015. Agglomeration economies in knowledge production over the industry life cycle: Evidence from the ICT industry in the Seoul capital area, south Korea. *International Journal of Urban Sciences* 19(3): 400–417.

Klepper, S. 1996. Entry, exit, growth, and innovation over the product life cycle. *American Economic Review* 86: 562–583.

Klepper, S. 1997. Industry life cycles. *Industrial and Corporate Change* 6: 145–182.

Korean Intellectual Property Office (KIPO). 2015a. The intellectual property statistics for 2014, Retrieved October 21, 2015, from http://www.kipo.go.kr/kpo/user.tdf?a=user.html.HtmlApp&c=3041&catmenu=m02_05_01.

Korean Intellectual Property Office (KIPO). 2015b. The table for matching up KSIC with IPC, Retrieved July 3, 2015, from http://www.kipo.go.kr/kpo/user.tdf?a=user.html.HtmlApp&c=4031&catmenu=m02_09_05_01.

Lim, U. 2004. Knowledge spillovers, agglomeration economies, and the geography of innovative activity: A spatial econometric analysis. *The Review of Regional Studies* 34(1): 11–36.

Mairesse, J., and P. Mohnen. 2002. Accounting for innovation and measuring innovativeness: An illustrative framework and an application. *The American Economic Review* 92(2): 226–230.

Mairesse, J., P. Mohnen, E. Kremp, and E. Kremp. 2005. The importance of R&D and innovation for productivity: A reexamination in light of the French innovation survey. *Annales d'Économie et de Statistique* 79/80: 487–527.
Mansfield, E. 1995. Academic research underlying industrial innovations: Sources, characteristics, and financing. *The Review of Economics and Statistics* 77(1): 55–65.
Markusen, A. 1996. Sticky places in slippery space: A typology of industrial districts. *Economic Geography* 72(3): 293–313.
McCann, P. 2007. Sketching out a model of innovation, face-to-face interaction and economic geography. *Spatial Economic Analysis* 2(2): 117–134.
Ministry of Employment and Labor (MOEL). 2015. *The labor demand survey*. Retrieved October 22, 2015, from http://kosis.kr/statHtml/statHtml.do?orgId=118&tblId=DT_118N_DEN021&conn_path=I3.
Nesta, L., and V. Mangematin. 2002. Industry life cycle, knowledge generation and technological networks. In *Networks, alliances and partnerships in the innovation process*, ed. F. Gault and J. de La Mothe, 199–220. New York: Springer.
OECD.stat. 2015. Business enterprise R-D expenditure by industry (ISIC 4). Retrieved October 19, 2015 from http://stats.oecd.org/Index.aspx?DataSetCode=BERD_INDUSTRY_ISIC4.
Park, I.K., and B. Von Rabenau. 2011. Disentangling agglomeration economies: Agents, sources, and spatial dependence. *Journal of Regional Science* 51(5): 897–930.
Polenske, K.R. 2001. Competitive advantage of regional internal and external supply chains. In *Regional science perspective in economic analysis*, ed. M.L. Lahr and R. E. Miller, 259–284. Amsterdam: Elsevier.
Porter, M.P. 1990. *The comparative advantage of nations*. New York: The Free Press.
Porter, M.E., and S. Stern. 1999. *New challenge to America's prosperity: findings from the innovation index*. Washington DC: Council on competitiveness.
Potter, A., and H.D. Watts. 2011. Evolutionary agglomeration theory: Increasing returns, diminishing returns, and the industry life cycle. *Journal of Economic Geography* 11(3): 417–455.
Solow, R. 1956. A contribution to the theory of economic growth. *The Quarterly Journal of Economics* 70(1): 65–94.
Sonn, J.W., and I.K. Park. 2011. The increasing importance of agglomeration economies hidden behind convergence: Geography of knowledge production. *Urban Studies* 48(10): 2180–2194.
Statistics Korea (KOSTAT). 2008. The table for matching up KSIC Rev. 9 with KSIC Rev. 8, Retrieved July 3, 2015, from https://kssc.kostat.go.kr:8443/ksscNew_web/index.jsp.
Statistics Korea (KOSTAT). 2015a. The Gross Regional Domestic Product (GRDP) by region, Retrieved October 19, 2015, from http://kosis.kr/statHtml/statHtml.do?orgId=101&tblId=DT_1C61&conn_path=I3.
Statistics Korea (KOSTAT). 2015b. The mining and manufacturing survey, Retrieved September 2, 2015, from http://kosis.kr/statHtml/statHtml.do?orgId=101&tblId= DT_1F01010&conn_path=I3.
Storper, M., and A.J. Venables. 2004. Buzz: Face-to-face contact and the urban economy. *Journal of Economic Geography* 4: 351–370.
Vernon, R. 1966. International investment and international trade in the product cycle. *Quarterly Journal of Economics* 80(2): 190–207.

Spatial Agglomeration and Firm Performance in Korean Manufacturing Industry, 2012

Ayoung Kim and Euijune Kim

Abstract This study tries to answer whether agglomeration economies lead to better firm performance or not. By adopting the random-intercept-multilevel model for 2012 Korean manufacturing data, we suggest an econometric specification strategy of the constant returns to scale (CRS) Cobb-Douglas production function in the multilevel structure, estimate the specified model, and analyze the results. Adopting two types of agglomeration economies represented by specialization and diversification, the results discussed in this paper can be summarized into three policy implications. First, specialization and diversification show the opposite effects on firm performances in most regions except the regions in some large metropolitan areas. In an ideal situation, both effects are not a trade-off phenomenon, and highly agglomerated cities are expected to have synergies from both effects. In the 2012 manufacturing sector in Korea, however, the offset between the two factors is observed. This means before the central and local governments implement industrial policy, they need to consider the existing mix of manufacturing sectors to not lose one of the agglomeration economies. Second, the specialization effect is relatively weaker than the diversification effect across regions. Even though there is no rule of proper effect size on both factors, these weak specialization effects can be seen as a big threat to the current economic growth strategies in Korea. If this specialization fails at a region level due to the weak specialization economies, the policy goal may not be achieved. Last, spatial heterogeneity in intercepts of the regional level dominates both specialization and diversification effects. In addition, diversification follows the trend of spatial heterogeneity. In 2012, the production performance of manufacturing firms leaned heavily on the region-specific factors not explained by the two agglomeration variables.

A. Kim
Department of Agricultural Economics, Purdue University, West Lafayette, IN, USA
e-mail: kim1426@purdue.edu

E. Kim (✉)
Department of Agricultural Economics and Rural Development and Research Institute of Agricultural and Life Science, Seoul National University, Seoul, Korea
e-mail: euijune@snu.ac.kr

E. Kim, B.H.S. Kim (eds.), *Quantitative Regional Economic and Environmental Analysis for Sustainability in Korea*, New Frontiers in Regional Science: Asian Perspectives 25, DOI 10.1007/978-981-10-0300-4_5

Considering the fact that there have been many policy concerns to resolve regional imbalance in economic growth, this questions the effectiveness of the previous efforts. From this standpoint, the strong spatial heterogeneity and the following trend of diversification emphasize that the local or central government, which tries to boost the economy in a lagging region and to achieve a well-balanced regional economy in a county, may want to think about the human capital or the other factors to increase productivity rather than just industry allocation strategy.

Keywords Agglomeration economies • Firm productivity • Multilevel model • Industrial location policy

1 Introduction

Firms and labor in highly agglomerated areas tend to be more productive than those in less-agglomerated areas. These economic externalities, which are caused by labor market pooling, input sharing, and knowledge spillovers (Marshallian channels) in a large and agglomerated market, are generally referred to as agglomeration economies. Agglomeration economies commonly lead to a regional economic growth process in urban areas and in industrial clusters by improving firms' performance. Since Marshall (1890) noted the advantages from the concentration of economic activities, agglomeration economic issues have been drawing attention from regional scientists and urban economists. Over the past 30 years, the advantages of agglomeration have been well documented and quantified, and, in particular, much of the empirical research in the agglomeration literature has investigated the effects of the spatial concentration of economic (or industrial) activities in terms of regional productivity or growth (Puga 2010).

Previous studies have generally measured the regional or industrial productivity gained from the spatial agglomeration of people, labor, firms, or industries. Most studies have been concerned with manufacturing industries which are well fitted to production function estimation with extensive annual and spatial data (Baldwin et al. 2008; Graham 2000; Henderson 1986, 2003; Lall et al. 2004; Moomaw 1981, 1983, 1985; Nakamura 1985; Sveikauskas et al. 1988; Tabuchi 1986). In addition, many studies of agglomeration use aggregate-level data from cities, counties, or metropolitan areas as the basic reference units, whereas few studies utilize firm-level data (Graham and Kim 2008; Henderson 2003; Mion and Naticchioni 2005; Andersson and Lööf 2011; Wheeler 2001). The firm-based data is accessible only with difficulty due to confidentiality, and thus previous studies have rarely dealt with the effect of agglomeration on firms' productivity.

The literature on agglomeration demonstrates that there are positive productivity gains from geographical clustering (Melo et al. 2009). However, several studies show that only localization economies—economies of scale arising from the spatial concentration of activity within industries—are of particular importance. On the other hand, urbanization economies—economies of scale arising from city size itself—despite their importance, have smaller effects on productivity (Rosenthal and Strange 2001). However, the results of empirical research are inconclusive (Beaudry and Schiffauerova 2009; de Groot et al. 2009; Puga 2010; Rosenthal and Strange 2004).

van Oort et al. (2012) point out that the ambiguity in the results of research on agglomeration economies is due to the missing link of the relationship between agglomeration economies and individual firm performance. They also show that the heterogeneity issue as it pertains to firm and region is a key element in terms of the scale of space, the level of aggregation, and the definition of growth. The relationship between agglomeration economies and regional growth is described as a macro (regional)-level phenomenon. However, the underlying theory of agglomeration is based on microeconomic foundations (firm-level) as well as a macro-level perspective (Duranton and Puga 2004; Rosenthal and Strange 2004) The micro-macro problem in the social sciences is also known as the "ecological fallacy" (Robinson 1950) or "cross-level fallacy" (Alker 1969), which is an error of deduction that involves deriving conclusions about individuals solely on the basis of an analysis of group data. Therefore, to examine agglomeration effects from a micro-level (firm-level) perspective, it is important to also understand the nature of agglomeration economies at the macro-level (regional-level) perspective.

Furthermore, different types of agglomeration may affect a firm's performance in various ways. A firm's activity or performance is strongly related to the regional economic environment, which can help explain why some firms are more productive than others. If individual firms are supposed to have the same attributes, being in the same region will tend to cause the performance of the firms to be more similar than they otherwise would be. In other words, once the grouping by spatial location has been established, even if the firms are randomly selected, they are affected by the homogeneous conditions of agglomeration in a region. Each regional group itself will tend to become differentiated from the others. This implies that the region and its firm members can both influence and be influenced by the composition of the group (Corrado and Fingleton 2012; Goldstein 1998). Therefore, multilevel (or hierarchical) modeling, which allows micro-levels and macro-levels simultaneously, is appropriate to analyze the relationship between spatial agglomeration and firm performance while addressing heterogeneity at the firm and regional levels. It also provides a practical tool to assess the extent to which a link exists between the macro-level (region) and the micro-level (firm).

In this paper, we start with a simple question: do agglomeration economies lead to better firm performance? In other words, which types of agglomeration economies give benefits to a firm's performance in the context of micro- and macro-level heterogeneity and interrelationships? This paper examines the link between external economies of localization (specialization) and urbanization (relative diversity) on firms' performance (productivity) in the multilevel framework. Thus, the hypothesis of this study is as follows: firms in highly agglomerated regions with a high level of diversity or specialization will have an advantage of productivity over firms in less-agglomerated regions.

The study has two aspects that separate our study from previous studies: (1) firm-level data is utilized to investigate the micro-foundation perspective in agglomeration economies and (2) multilevel modeling, which includes different levels of data (e.g., firm-level and regional-level data), is applied to explain the effects of micro- and macro-level heterogeneity on firm performance. The paper is structured as follows: in the next section, we review multilevel modeling, the following section explains the empirical model specification in a spatial framework, and Sect. 4 describes the variables from firm-level and regional aggregate-level data utilized in this empirical study.

2 Multilevel Modeling

Spatial perspective analysis generally utilizes either cross-sectional data indexed by location rather than time or panel data in which each time layer includes a data set of location-specific observations. In an economic analysis with a hierarchy of local, regional, and national levels, the spatial level affects outcomes. Therefore, multilevel modeling is a good starting point to deal with a micro-macro problem in which an individual level of cross-sectional data is constructed within the same region (Corrado and Fingleton 2012). For instance, the levels of agglomeration are different across spatial (geographical) units (e.g., local administrative areas), and the properties of agglomeration partly reflect the observed performance of a firm in the area. Figure 1 provides a diagrammatic representation of a multilevel hierarchy. It consists of two levels: the top level, R, denotes the regional level (local administrative area, i.e., municipal level, Si/Gun/Gu in Korea), and F indicates the individual establishment level. There is a variable number of firms in each category at the regional level. In the context of spatial data, R represents a geographical grouping of firms.

Firstly, a simple linear regression model is introduced to analyze a firm's production function using firm-level data and regional characteristics represented by industrial specialization and relative diversity. For the$j$th region, the model is defined as follows:

$$\mathbf{Y}_j = \mathbf{X}_j\beta_j + \mathbf{z}_j\gamma_j + \varepsilon_j \tag{1}$$

where $\mathbf{Y}_j = \left\{y_{ij}\right\}$ denotes a $(F_j \times 1)$ vector of an individual firm's output (value added); $\mathbf{X}_j = \left\{x_{ij}\right\}$ is a $(F_j \times (k+1))$ covariate matrix of firm-level inputs (labor and capital) including the intercept term; $\mathbf{z}_j$ is a $(F_j \times q)$ matrix of regional-level explanatory variables (the level of regional agglomeration, the level of specialization and diversity) and is invariant within regions; $\varepsilon_j = \left\{e_{ij}\right\} + \left\{u_j\right\}$ is the random disturbance vector with e_{ij}or the i unit belonging to level j and a random effect u_j accounting for some level two heterogeneity; i $\left(i = 1, \ldots, F_j\right)$ and j $(j = 1, \ldots, R)$ denote individual-level (firm-level) and regional-level units, respectively; and γ_j denotes the $(q \times 1)$ vector of fixed effect coefficients.

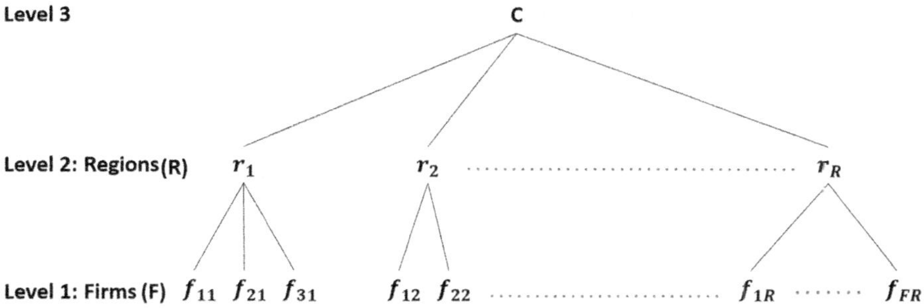

Fig. 1 A hierarchical structure (modified from Corrado and Fingleton (2012))

For a simple approach to estimate the data with nested groups of observations, the standard uni-level OLS estimation can be used with dummy variables for the region. However, a weakness of this approach is that the large number of levels (i.e., large number of regions, in this paper 263 for Si/Gun/Gu) results in a dramatic reduction in degrees of freedom. In addition, in assessing any causal relationships between y_{ij} and one or more elements of x_{ij}, it is necessary to consider the hierarchical structure of the data and in particular the within- and between-group effects. The multilevel approach helps to analyze the effect of heterogeneous groups. In fact, with unbalanced data, while OLS estimates of the coefficients give equal weights to each cluster, the variance component model acknowledges the fact that the estimates for the fixed coefficients can change according to the cluster size. Applying the multilevel model instead of the simple linear model of Eq. (1) can manage the heterogeneity of the firm and regional level simultaneously. By including agglomeration variables represented by $\mathbf{z}_j$ in an extension of Eq. (1), the multilevel structure is set up as follows:

$$\mathbf{Y}_j = \mathbf{X}_j\beta_j + \varepsilon_j \tag{2}$$

$$\beta_j^k = \gamma_0^k + z_j'\gamma_j^k + u_j^k \tag{3}$$

where $\mathbf{Y}_j = \{y_{ij}\}$ denotes a $(F_j \times 1)$ vector of an individual firm's output (value added), $\mathbf{X}_j = \{x_{ij}\}$ is a $(F_j \times 1)$ covariate matrix of firm-level inputs (labor and capital stock), $\mathbf{z}_j'$ is a $(q \times 1)$ vector of regional-level agglomeration and is invariant within regions, $\varepsilon_j = \{e_{ij}\} + \{u_j\}$ is the random disturbance, β_j^k denotes the k coefficient of the firm-level model, and γ_j^k represents the regional-level coefficient of the kth firm-level coefficient in the j th region.

3 Empirical Model Specification

We need to specify the general form of multilevel model in Eqs. (2) and (3) as a reduced form equation. In economic literature, the Cobb-Douglas production technology is generally adopted to analyze firm's production behavior. To make a comparison of individual firms' performance, we assume the constant returns to scale (CRS) and specify an empirical model of Eq. (2) as a CRS Cobb-Douglas function with a firm i in a region j as

$$y_{ij} = \alpha_0 L_{ij}{}^{\beta} K_{ij}{}^{1-\beta} + \varepsilon_{ij}, \tag{4}$$

where y_{ij} is output, L_{ij} is labor input, K_{ij} is capital stock, α_0 is technological coefficient, β is the elasticity coefficient of labor, and ε_{ij} is random disturbance term. For the purpose of estimation, we take the logarithms on both sides and derive the following equation:

$$\log y_{ij} = \alpha + \beta \log L_{ij} + (1 - \beta) \log K_{ij} + \epsilon_{ij}, \tag{5}$$

where $\alpha = \log\alpha_0$ and $\epsilon_{ij} = \log\varepsilon_{ij}$. Following the conventional assumption, we assume that ϵ_{ij} is *i.i.d.* normal. Equation (5) is a classic and generally adopted approach to estimate production coefficients. In an estimation process of multilevel modeling, however, assigning the CRS restriction is not usually applicable in many econometric software. By taking additional calculation on Eq. (5), we can apply the CRS restriction in the form available in general software packages as

$$\log\left(y_{ij}\right) - \log\left(K_{ij}\right) = \alpha + \beta\left(\log L_{ij} - \log K_{ij}\right) + \epsilon_{ij}. \tag{6}$$

The elasticity coefficient of capital can be derived from $(1 - \beta)$ as given in the CRS assumption after estimating β.

Now, we need to describe how to specify the regional-level structure to the multilevel models of Eq. (3). In urban economic literature, individual firms produce outputs from their inputs, and they are getting regional advantages, i.e., specialization and diversification benefits of agglomeration from the region where they are located. In model specification to find the effect of agglomeration, we need to know which parameters (or variables) reflect the agglomeration economies in Eq. (6). In this paper, we assume that two major advantages come from the technological coefficient, α, and the labor elasticity coefficient of β solely taken by labor inputs. The data detailed in the next section is a single cross section. Within a year, therefore, a firm's adoption of input combinations in their production technology is not only allowable in labor but in capital, i.e., the short-term production portfolio without uncertainty is assumed. Under the given circumstance, firms are assumed to completely know their production frontiers and achieve the optimal production level. As a consequence, they include all the regional advantages of agglomeration into their non-input factors to reduce their production costs. In the CRS Cobb-Douglas production technology under single-year production data, the only available factor to reflect this assumption is α.[1] The model we adopt is, therefore, a random intercept model as

$$\log\left(y_{ij}\right) - \log\left(K_{ij}\right) = \alpha_j + \beta\left(\log L_{ij} - \log K_{ij}\right) + \epsilon_{ij}, \quad (7)$$

$$\alpha_j = \mathbf{z}_j^{'}\gamma_j + u_j^k. \quad (8)$$

As notated in Eq. (7), α_j has the regional subscript j from Eq. (6). The second-level equation comes with the regional-level variables (intercept and two variables described in the next section, specialization and diversification) in Eq. (8). In the estimation, we adopt the restricted maximum likelihood estimation (REML) (Goldstein 2011; Kreft and de Leeuw 2002) and "*lme4*" R packages are applied.

4 Data

In this analysis we utilize data based on the plant level underlying the Annual Report on Mining and Manufacturing Survey of Korea (2012) by the Korea National Statistical Office. The survey covers all plants with five or more employees in mining and manufacturing industries and contains information on the production process that is necessary to estimate plant-level production function. The data are classified into three administrative area levels: province (Do/Gwangyeoksi), municipal (Si/Gun/Gu), and submunicipal (Eup/Myeon/Dong/Ri). However, the survey only provides the individual data for the province and municipal levels.

This paper focuses on 23 manufacturing subsectors for a single year, 2012. The key variables to estimate firm-level production functions are labor, capital stock and value added with industry classification, and location information at various geographical levels, including province and municipal.[2] Thus, the municipal is the reference unit for the regional level in this analysis. This paper presents an investigation into the connection between a firm's productivity and the externalities from agglomeration including diversity and specialization. Agglomeration economies are assumed to be club goods that operate at a large regional scale. The levels of agglomeration as regional attributes in this paper are also proxy measures identical to the municipal level.

[1] Depending upon assumptions applied to model specification, the agglomeration factors can be applied to any coefficient in the assigned production technology. For model simplicity and data availability, we stay in flexible α only.

[2] Location information at the submunicipal level is not provided in the survey due to the disclosure policy.

The benefits of the spatial concentration of economic activities can be earned from the intra- and interindustry clustering of economic activities, referred to as urbanization and localization economies in the agglomeration literature (Melo et al. 2009). These economies can be measured by specialization and diversity indices. Beaudry and Schiffauerova (2009) summarize the indices used for localization and urbanization in the empirical studies that investigate the effect of agglomeration economies.

Localization economies are generated when many firms or jobs in the same industry are located close to each other. Regions whose production is a structure specialized in a particular industry tend to be more productive in that particular industry, as this arrangement allows knowledge to spill over between similar firms. The most common measurement of localization economies is the location quotient (LQ) introduced by Florence (1939), which is often employed to quantify industrial concentration in regions. Typically, LQ is constructed with employment data, and the measure is the ratio of two shares: the employment share of a particular industry in a region and the employment share of that industry in a wider area such as a country. Researchers often assume that if the quotient is above one, then the industry is concentrated in the region. The LQ to assess industry localization in a particular region is defined as follows:

$$LQ_{ij} = \left(\frac{E_{ij}}{E_j}\right) / \left(\frac{E_{in}}{E_n}\right)$$

where E_{ij} is employment in industry sector i and region j; E_n or E_j is total employment at the national level or regional level, respectively; and E_{in} is the total national employment in industry sector i. Alternatively, own-industry employment can be used to measure localization economies.

Urbanization economies are created when firms cross their industrial boundaries; knowledge spills over between different industries, causing diversified production structures to be more productive (van der Panne 2004). The diversity-based (inverse) Herfindahl-Hirschman index (HHI) is the most used measure for urbanization economies. There are a few alternatives to inverse HHI to measure urbanization economies, such as the Gini index, the total local population, the total local employment, or other industry employment. In this paper we utilize the relative-diversity index by summing for each region, overall sectors, the absolute value of the difference between each sector's share in local employment and its share in national employment (Duranton and Puga 2000). The relative-diversity index is given by

$$RDI_j = 1/\sum_j \left|s_{ij} - s_i\right|$$

where s_{ij} is the share of industry i in region j and s_i is the share of industry i in national employment.

Table 1 Descriptive statistics in manufacturing industry (n=53,045)

Variable (2012)	Mean	Min	Max	Std. dev.
Value added	7090	1.00	9,515,833	103,977
Employment	43.04	1.00	26,877	249.25
Capital stock	7764	1.00	9,893,439	113,031
Location quotient	3.99	0.01	958.34	11.61
Relative-diversity	0.035	0.01	0.49	0.04
Number of firms	53,045			

Table 1 presents descriptive statistics for the data which is utilized in our analysis. The original survey includes 65,743 firms' production data in the mining and manufacturing industries, but we only focus on the manufacturing industry. Due to the disclosure policy,[3] available observations are 53,045. Due to the logarithmic transformation in the specified model of Eqs. (7) and (8), any observations with values of zero are changed to one following the conventional rule of thumb.

5 Results

We first estimate the Cobb-Douglas production function of Eq. (6) to investigate regional-level variation in residuals. By adopting the firm-level data, the OLS results are presented in Table 2 based on firm-level observations.

As noted in the previous section, Eq. (6) is assumed to be the CRS Cobb-Douglas production technology, and the elasticity of capital is derived from the estimated elasticity of labor. As generally expected, the adopted data shows that the outputs are more heavily dependent on labor inputs (0.826) and intercepts contributed about 3.612 to the variations of outputs.

To test whether there are regional-level effects on the productivity, the residual from the OLS results should be considered and are summarized by regional level (Si/Gun/Gu) in Fig. 2. Each boxplot represents the distributional characteristics of residuals by each region.

As shown in Fig. 2, the residual boxplots vary across regions greatly, particularly from the standpoint of quartiles. Even though it's a good model fit in the results of Table 2 (R-squared = 84.6 %), the OLS estimation does not fully explain the variation of individual firm-level performances. As we argued, we tackle two regional-level factors—specialization and diversification—into Eqs. (7) and (8). We estimate the random intercept model and the results are shown in Table 3.

[3] Korea Statistical Office adopts a disclosure rule to protect individual firm's information. About 19.31% (12,698) of firm observations are masked, and 53,045 in total is the maximum availability for this study.

Table 2 OLS result of production function at firm level

Variable	Coefficients
Intercept	3.612***
	(0.0061)
Ln(employment)	0.826***
	(0.0015)
Ln(capital) =1 − Ln(employment)	0.175
Number of obs.	53,045
R^2	84.6

Note: significance levels of 10 %, 5 %, and 1 % are denoted by *, **, and ***, respectively

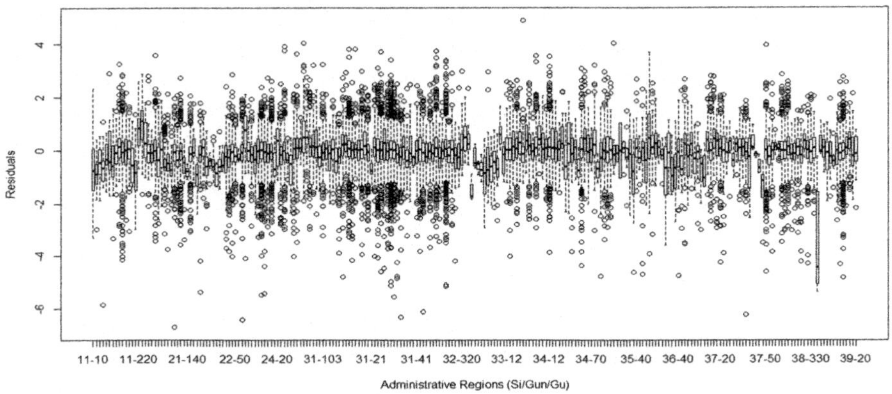

Fig. 2 Residual plot for production function at administrative level (Si/Gun/Gu)

Table 3 Multilevel results for firm production function

Fixed effects		Random effects				
			Variance	Std. dev.	Corr.	
Intercept	3.641*** (0.015)	Intercept	0.023	0.151		
Ln(employment)	0.834*** (0.002)	Specialization	0.001	0.034	−0.12	
Ln(capital) = 1 − Ln (capital)	0.166	Diversification	0.192	0.438	0.19	−0.49
Number of firms	53,045	Log likelihood	−53703.44			

Note: significance levels of 10 %, 5 %, and 1 % are denoted by *, **, and ***, respectively

In Table 3, the estimates are statistically significant with 1 % significance level, and the size of estimated labor elasticity is increased to 0.834. From the negative correlation between specialization and diversification, we can expect that there is a trade-off relation between two regional-level factors. From the larger variance of diversification, we can infer that diversification is a key driver of regional-level

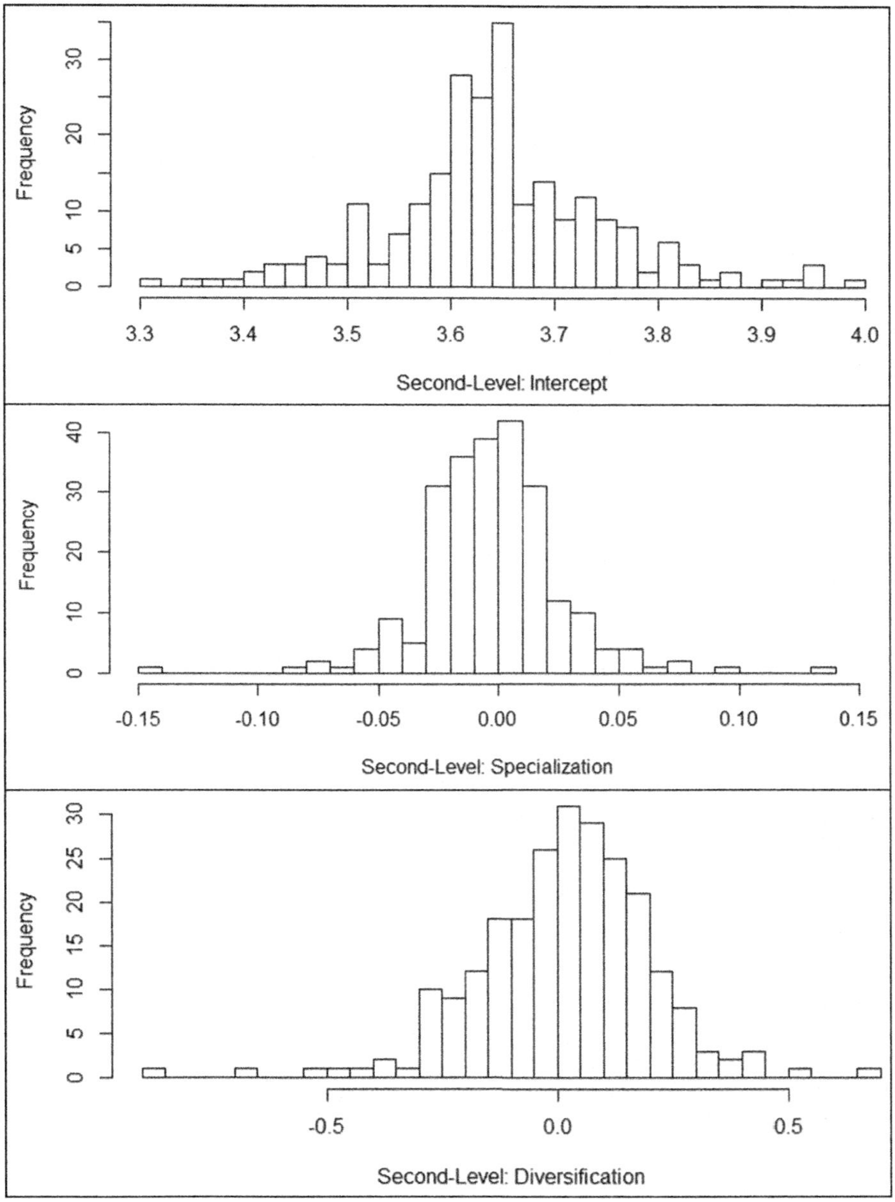

Fig. 3 Histogram of coefficients in second-level model

difference in agglomeration economies on the production performances of individual firms.

To analyze the distribution of the second-level coefficients, we construct the histograms of regional-level estimates in Fig. 3.

The second and third panels of Fig. 3 are the distribution of specialization and diversification estimates at the regional level. The interesting fact is that the specialization estimates are right skewed while the diversification estimates are left skewed. As described above, this explains the opposite directional movement of two effects. The highly specialized regions are likely to have lower diversification and vice versa.

The first panel in Fig. 3 is the distribution of intercepts, and they are all positive values and dominate the other two in the point of magnitude. This means there exist some region-specific positive production factors not captured by the two variables of agglomeration economies. Considering the relatively smaller magnitude of measurement units in specialization and diversification, the larger magnitude of intercept itself does not necessarily mean larger spatial heterogeneity. It is, however, noteworthy that region-specific economic factors and conditions are important elements which need to be considered in local-level economic growth policy. To further study the trade-off relation, we plot the regional-level estimates followed by the size of intercepts in Fig. 4.

In Table 3, the given variation of specialization (0.001) is the smallest while that of diversification (0.192) is the largest. Figure 4 shows the exactly matched vibrational bandwidth in the second and third columns. While specialization estimates do not vary across the changes of intercepts, the estimates of diversification vary as the similar shape of slope in the estimates of intercepts. This implies that diversification could be an endowed factor in each region. Considering the fact that the production performance of individual firms can be seen as highly dependent upon the built-in environment in a region, the level of diversity by industries or firms in a region significantly influences firm's productivity due to larger variation in diversification. On the other hand, the smaller variation of magnitude of specialization is not a strong leading factor for a firm's production. This implies that the local government may encounter limited resources and flexibility in its economic policy.

To empirically confirm this argument, we plot the second-level coefficients on the map of municipal level (Si/Gun/Gu) as shown in Fig. 5. In Fig. 5, the mean-deviated estimates are plotted by each region. The regions having no manufacturing sectors are presented as empty color. The regions in darker red (blue) color present the above (below) average effect of variables on firm productivity. The impacts of specialization and diversification vary across the region due to the different industrial structure which already exists.

As expected, specialization and diversification have the opposite direction of color patterns, i.e., the higher specialization means the smaller diversification and vice versa. In addition, intercepts and diversification present the similar color pattern as described above. It is obvious that Seoul metropolitan area shows a larger impact due to diversification and specialization, but is around the average

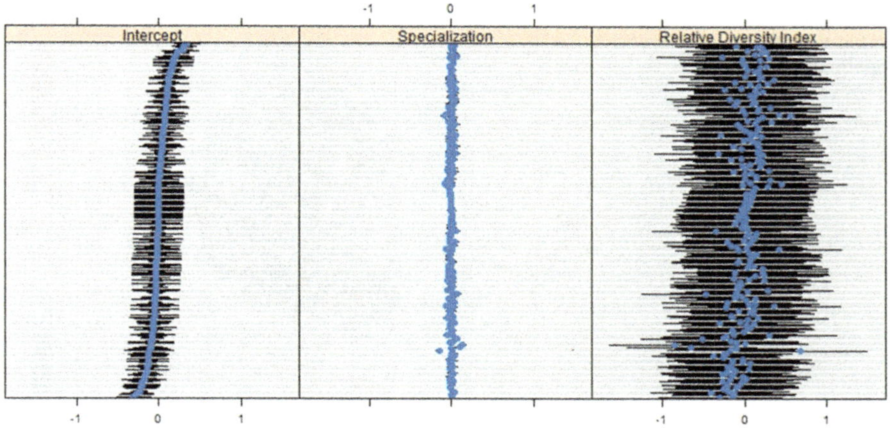

Fig. 4 Random effect variation

(mean of coefficient in specialization is -0.003 and mean of coefficient in diversification is 0.014). The regions with industrial complexes built up by the central government in the 1970s, for example, Ulsan, Pohang, and major heavy manufacturing industry-based cities, also present a moderate size of specialization and diversification effects. However, the regions along the Gyeongbu Expressway show higher values (above average) for the impact of specialization and lower values (below average) for the impact of diversity. For those regions, clustering manufacturing industries or firms lead to increasing a firm's productivity.

6 Conclusion

This paper tries to answer whether agglomeration economies lead to better firm performance or not. By adopting the random-intercept-multilevel models of 2012 Korean manufacturing data, we suggest an econometric specification strategy of the CRS Cobb-Douglas production function in the multilevel structure, estimate the specified model, and analyze the results. Adopting two regional-level agglomeration economies represented by specialization and diversification, the results discussed in this paper can be summarized into three policy implications.

First, specialization and diversification show the opposite effects on firm performances in most of the regions except the regions in some large metropolitan areas. In an ideal situation, both effects are not a trade-off phenomenon, and highly agglomerated cities are expected to have synergies from both effects. In the 2012 manufacturing sector in Korea, however, the offset between these two factors is observed. This means before the central and local governments implement industrial policy, they need to consider the existing mix of manufacturing sectors to not lose one of the agglomeration economies.

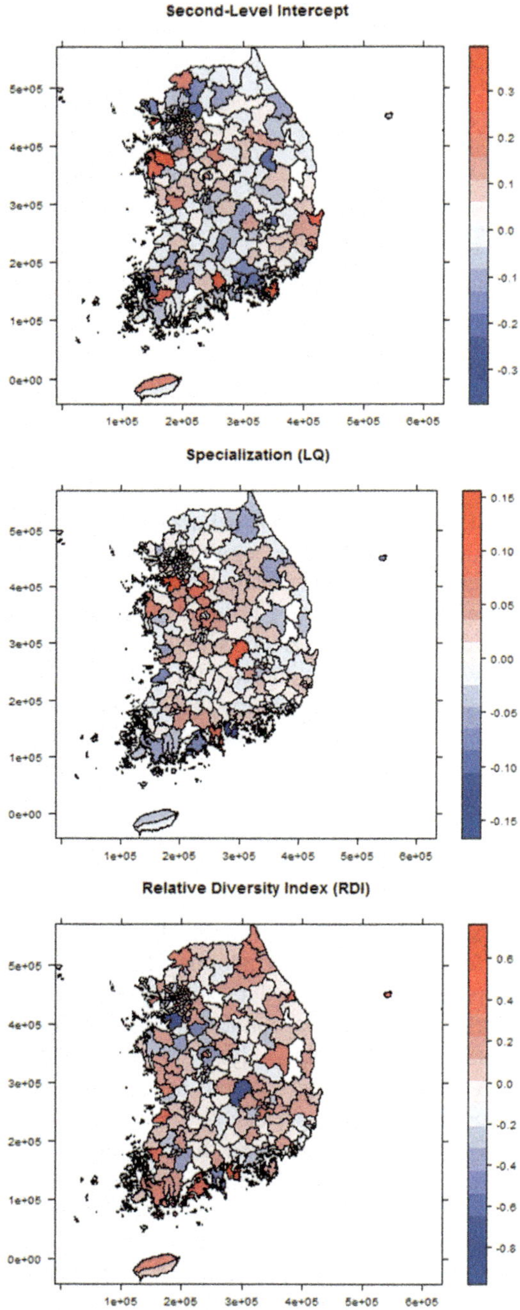

Fig. 5 Map plots of the mean-deviated second-level coefficients

Second, the specialization effect on a firm's productivity is relatively weaker than diversification across regions. Even though there is no rule of proper effect size on both factors, this weak specialization effect can be seen as a big threat to the current economic growth strategies in Korea. Both local and central governments emphasize the role of new specialized industry complexes in a region, i.e., building up a new technology or IT industry concentrating on a city and expecting to link it to other cities for a nationwide diversification. If this specialization fails in a region level due to the weak specialization economies, the policy goal may not be achieved. The imperative and urgent policy is, therefore, to strengthen specialization effects at a regional level. Then, diversification tends to follow up the effects as given in the current status.

Last, spatial heterogeneity in intercepts of the regional level dominates both specialization and diversification. In addition, diversification follows the trend of spatial heterogeneity. From 2012 manufacturing data, the production performance of manufacturing firms leaned highly on the region-specific factors not explained by the two agglomeration variables. Regional attributes such as the density of human capital, the level of infrastructure capacity, or the proportion of high-tech manufacturing firms can strongly influence firms' productivity. Considering the fact that there have been many policy concerns to resolve regional imbalance in economic growth, this questions the effectiveness of the previous efforts. From this standpoint, the strong spatial heterogeneity and the following trend of diversification emphasize that the local or central government, which tries to boost the economy in a lagging region and to achieve a well-balanced regional economy in a county, may want to think about the human capital or the other factors to increase productivity rather than just industry allocation strategy.

References

Alker Jr., H. 1969. A typology of ecological fallacies. In *Quantitative ecological analysis in the social sciences*, ed. M. Dogan and S. Rokkan. Cambridge: MIT Press.

Andersson, M., and H. Lööf. 2011. Agglomeration and productivity: Evidence from firm-level data. *The Annals of Regional Science* 46: 601–620.

Baldwin, J., D. Beckstead, W. Mark Brown, and D. Rigby. 2008. Agglomeration and the geography of localization economies in Canada. *Regional Studies* 42: 117–132.

Beaudry, C., and A. Schiffauerova. 2009. Who's right, Marshall or Jacobs? The localization versus urbanization debate. *Research Policy* 39: 318–337.

Corrado, L., and B. Fingleton. 2012. Where is the economics in spatial econometrics? *Journal of Regional Science* 52: 210–239.

de Groot, H., H. Poot, and M. Smit. 2009. Agglomeration externalities, innovation and regional growth: Theoretical perspectives and meta-analysis. In *Handbook of regional growth and development theories*, ed. P. Nijkamp and R. Capello. Cheltenham: Edward Elgar.

Duranton, G., and D. Puga. 2000. Diversity and specialization in cities: Why, where and when does it matter. *Urban Studies* 37(3): 533–555.

Duranton, G., and D. Puga. 2004. Micro-foundations of urban agglomeration economies. In *Handbook of regional and urban economics*, vol. 4, ed. J. Henderson and J. Thisse. Amsterdam: Elsevier.

Florence, P. 1939. *Political and economic planning. Report on the location of industry. A survey of present trends in Great Britain affecting industrial location and regional economic development with proposals for future policy*. London: PEP.
Goldstein, H. 1998. *Multilevel models for analysing social data*. Technical report. Encyclopedia of Social Research Methods.
Goldstein, H. 2011. *Multilevel statistical models*, vol. 922. Hoboken: Wiley.
Graham, D. 2000. International review of applied economics. *Journal of Regional Science* 14: 323–340.
Graham, D., and H. Kim. 2008. An empirical analytical framework for agglomeration economies. *The Annals of Regional Science* 42: 267–289.
Henderson, J. 1986. Efficiency of resource usage and city size. *Journal of Urban Economics* 19: 47–70.
Henderson, J. 2003. Marshalls scale economies. *Journal of Urban Economics* 53: 1–28.
Kreft, I., and J. de Leeuw. 2002. *Introducing multilevel modeling*. London: Sage.
Lall, S., Z. Shalizi, and U. Deichmann. 2004. Agglomeration economies and productivity in Indian industry. *Journal of Development Economics* 73: 643–673.
Marshall, A. 1890. *Principal of economics*. London: Macmillan and Co., Ltd.
Melo, P., D. Graham, and R. Noland. 2009. A meta-analysis of estimates of urban agglomeration economies. *Regional Science and Urban Economics* 39: 332–342.
Mion, G., and P. Naticchioni. 2005. *Urbanization externalities, market potential and spatial sorting of skills and firms*, Technical report, CEPR, Discussion Papers.
Moomaw, R. 1981. Productivity and city size? A critique of the evidence. *Quarterly Journal of Economics* 96: 675–688.
Moomaw, R. 1983. Is population scale a worthless surrogate for business agglomeration economies? *Regional Science Urban Economics* 13: 525–545.
Moomaw, R. 1985. Firm location and city size: Reduced productivity advantages as a factor in the decline of manufacturing in urban areas. *Journal of Urban Economics* 17: 73–89.
Nakamura, R. 1985. Agglomeration economies in urban manufacturing industries: A case of Japanese cities. *Journal of Urban Economics* 17: 108–124.
Puga, D. 2010. The magnitude and causes of agglomeration economies. *Journal of Regional Science* 50: 203–219.
Robinson, W.S. 1950. Ecological correlations and the behavior of individuals. *American Sociological Review* 15: 351–357.
Rosenthal, S., and W. Strange. 2001. The determinants of agglomeration. *Journal of Urban Economics* 50: 191–229.
Rosenthal, S.S., and W.C. Strange. 2004. Evidence on the nature and sources of agglomeration economies. In *Handbook of regional and urban economics*, vol. 4, ed. J. Henderson and J. Thisse. Amsterdam: Elsevier.
Sveikauskas, L., J. Gowdy, and M. Funk. 1988. Urban productivity: City size or industry size. *Journal of Regional Science* 28: 185–202.
Tabuchi, T. 1986. Urban agglomeration, capital augmenting technology, and labor market equilibrium. *Journal of Urban Economics* 20: 211–228.
van der Panne, G. 2004. Agglomeration externalities: Marshall versus Jacobs. *Journal of Evolutionary Economics* 14: 593–604.
van Oort, F.G., M.J. Burger, J. Knoben, and O. Raspe. 2012. Multilevel approaches and the firm-agglomeration ambiguity in economic growth studies. *Journal of Economic Surveys* 26: 468–491.
Wheeler, C. 2001. Search, sorting, and urban agglomeration. *Journal of Labor Economic* 19: 879–899.

Impact of Local Government Monetary and Fiscal Policies on Output Growth of Firms

Changkeun Lee and Euijune Kim

Abstract This study explores the impacts of the monetary and fiscal policies of local governments on the output growth of firms, using multilevel statistical models. It concludes that short-term loans and bonds have a positive influence on output growth but long-term loans have a negative effect, while the effect of paid-in-capital increases is inconclusive and insignificant. A credit guarantee from the local government has the largest impact on the elasticity values of factor inputs and financing amounts with respect to output, compared to other government expenditure programs.

Keywords Monetary and fiscal policies • Multilevel statistical models • Bank-based and market-based financial systems • Local government • Financing options

1 Introduction

South Korea has experienced two major financial crises in the last two decades. In the midst of these shocks, cash-strapped firms have been in an ongoing restructuring and debt readjustment processes through the Korean Asset Management Company's Non-Performing Loan Disposal Fund and Structural Regulation Fund.[1] Under the restructuring process, the central government has implemented actively monetary policies (e.g., the extension of the loan, timely supply of funds for corporates, and purchasing speculative-grade bonds for prompt liquidation) and

[1] In 1997, the Korean government established the Corporate Restructuring Committee and instituted corporate restructuring emergency funding. In 2008, the Coordinating Committee of Creditor Financial Institutions was launched to advance the restructuring process.

C. Lee (✉)
Research Institute of Advanced Materials, Seoul National University, Seoul, Korea
e-mail: bigtwo7@snu.ac.kr

E. Kim
Department of Agricultural Economics and Rural Development and Research Institute of Agricultural and Life Science, Seoul National University, Seoul, Korea
e-mail: euijune@snu.ac.kr

E. Kim, B.H.S. Kim (eds.), *Quantitative Regional Economic and Environmental Analysis for Sustainability in Korea*, New Frontiers in Regional Science: Asian Perspectives 25, DOI 10.1007/978-981-10-0300-4_6

fiscal policies such as corporate tax integration and the introduction of a corporate cash flow tax. Besides, local governments have developed credit guarantee schemes of regional firms and have executed financing, taxation, and budgetary adjustments of early public finance expenditures, innovative activity assistance, and industrial policy regulations for entrepreneurial activities. However, it is not clear which financial and fiscal tools contribute most to a company's business and output growth.

The purpose of this study is to analyze the impact of monetary and fiscal policies of local governments on the output growth of firms. In this paper, we take into account four financing options for firms (a short-term loan, a long-term loan, a bond, and a paid-in-capital increase) and four monetary and fiscal policy options for the government to implement—a credit guarantee, a local tax revenue (or tax reduction), a non-investment expenditure, and an investment expenditure. While many works have focused on the economic role of the financial sector in economic growth, this paper explores how the output growth of firms depends on monetary and fiscal policies of local governments. This study estimates firms' output function using a multilevel structure of micro firm-level data and regional macro data from 1997 to 2009. The next two sections review the literature on the relationship among the financial sector, monetary and fiscal policies, and regional economic growth and analyze the effect of monetary and fiscal policies using multilevel statistical models. The final section summarizes the main findings and discusses future research topics.

2 Literature Review

2.1 Financial Systems[2] and Regional Economic Development

Beare (1976) was one of the pioneering works contributing to the development of monetary theory at the regional level. He argued, using a simple reduced form equation, that a national money and banking system controlled by a central bank was important in determining regional income. Rioja and Valey (2004), using dynamic panel generalized method of moments (GMM) techniques, found that the relationship between a country's financial development and its economic growth varies according to the development level of financial systems. For example, improvement in financial markets had an uncertain effect on economic growth at a low level of financial development but a large and positive effect on growth at an intermediate level. The effect was positive but smaller at an advanced level.

[2] The bank-based system emphasizes the role of banks in mobilizing capital and managing risk, while the market-based system emphasizes the role of markets in enhancing risk management, information dissemination, corporate control, and capital allocation (Lee 2012; Levine and Zervos 1998).

However, Rousseau and Wachtel (2002), using a core cross-country panel analysis, concluded that the relationship between financial systems and economic growth was unstable and depended on the types of regional and economic structures.

The discussion of the comparative advantages of bank-based and market-based financial systems originated with Gerschenkron's (1962) perspective on financial structures. Since the early 1980s, three issues have been raised regarding (1) the comparative advantage of the market-based system, (2) the dominant position of the bank-based system, and (3) the complementary relationship between these two financial systems. After the recession in the Japanese economy and the regression in the German economy in the 1990s, the financial market placed more emphasis on the market-based financial system than the bank-based one. Some have argued that the market-based financial system was superior in terms of efficient resource allocation (e.g., Porter 1992; Franks and Mayer 1993; Boyd and Gertler 1994; and Levine and Zervos 1998). Rajan and Zingales (1998a) explained that the bank-based financial system resulted in inefficient fund allocation because the system entailed oblique dealings, which might have caused the financial crisis in East Asia in 1997. Shin and Kim (2011) reported that, after the financial crisis of 1997, the profitability of banks was not enhanced solely through improvements in technological efficiency in Korea. Their profitability was improved through a restructuring process and through increased scale efficiency, as the banks struggled to increase their scale efficiencies to survive. This finding implies that such a scenario might lead to risk-averse attitudes on the part of banks.

However, the bank-based financial system has comparative advantages in financial intermediation through information gathering, valuation of companies, and maintaining relationships with companies (e.g., Gerschenkron 1962; Opler 1993; and Rajan and Zingales 1998b). Becketti and Morris (1992) and Hooks and Opler (1993) credited the contribution of the banking sector with the growth of small- or medium-sized companies, despite the development of the market-based financial system. Schmidt et al. (1999) found, based on the theory of financial intermediation, that there has been no general trend toward disintermediation, toward a transformation from the bank-based to the market-based financial system, or toward a loss of importance of banks, in three major European economies: France, Germany, and the UK. Stulz (2000) argued that banks are more effective at providing external resources to new and innovative activities that require staged financing because banks can credibly commit to making additional funding available as a project develops.

In analyzing the financial sector in 45 countries from 1980 to 1995, Levine (2002) concluded that a better developed financial system (between bank-based or market-based financial systems) induced economic growth. Beck and Levine (2001) reported that banks and stock markets are complementary: they jointly contribute to economic growth. They also found that distinguishing countries by overall financial development and legal system efficiency is more useful than distinguishing countries by whether they are relatively bank based or market based. Deidda and Fattouh (2008) showed theoretically, through a simple model of competitive financial systems, that both bank and stock market development have a positive effect on growth, but the growth speed of bank development is low when there is a high level of stock market development. Lee (2012) demonstrated, using Granger-causality tests, that while the banking sector plays an important role in the early stage of economic growth, recently the stock market has played a very substantial role in the economic growth of economically advanced countries (e.g., the USA, the UK, Germany, France, and Japan). He concluded that the banking sector and the stock market in each country were complementary to one another in fostering the process of economic growth.

2.2 Monetary and Fiscal Policies

Local government policies regarding the overall output growth of firms can be divided into three types: monetary policies (e.g., subsidies and credit offering), fiscal policies (e.g., taxation and public expenditures), and industrial policies (Harrison et al. 2004). To minimize the financial constraints on firms, especially small- and medium-sized firms, local and central governments employ diverse monetary and fiscal policies such as mutual funds, credit assistance, credit guarantees, venture capital, subsidies, various expenditures, and tax systems. Empirical evidence has revealed inconclusive results regarding the effectiveness of such monetary policies (see Table 1). Boocock and Shariff (2005), looking at the case of Malaysia, evaluated the credit guarantee scheme in terms of financial additionality (i.e., an increase in accessibility to loans) and economic additionality (i.e., utilizing funds to benefit firms and to generate positive spillovers). They concluded that the Malaysian credit guarantee scheme in general failed to satisfy financial additionality, although it showed some positive outcomes in relation to the economic additionality. Oh et al. (2009), using propensity score matching, suggested that credit guarantees significantly influence firms' ability to maintain their size in terms of sales and employment, and increase their survival rate, but they do not increase firms' R&D and investment or their growth in productivity.

Table 1 Impacts of monetary policies on the firm's growth

Author	Type of monetary policies	Impacts on firm's growth
Cowling and Mitchell (1997)	Credit guarantee program	Positive
Li (2002)	Credit assistance program	Positive
Lerner (2002)	Public loan support program	Negative
Meager et al. (2003)	Credit scheme	Negative
Kreft and Sobel (2005)	Venture funds	Negative
Boocock and Shariff (2005)	Credit guarantee scheme	Positive, but restrictive
Cumming (2007)	Venture capital	Positive
Oh et al. (2009)	Credit guarantee program	Positive, but restrictive

Fiscal policies, such as taxation and local and central government expenditures, can either encourage or discourage entrepreneurs seeking to establish businesses. For example, taxation exerts mainly a negative effect on firms' activities, but McGreal et al. (2002) demonstrated that taxation instruments were effective in achieving urban regeneration in Dublin (Ireland) and Chicago (USA). Local and central government expenditures increased output per capita and firm profitability and encouraged innovation (Jacobides et al. 2006; Chen and Groenewold 2011; Yu and Rickman 2013). Takii (2008) stressed that an expansionary fiscal policy can strengthen the social role of firms in predicting idiosyncratic changes in consumers' taste preferences if the government expenditure can reflect changes in consumer tastes. While such fiscal policies affect entrepreneurial activities and output growth, Takii (2008) came up with mixed results regarding their effectiveness.

3 Analysis

3.1 Data

This work uses a total of 3045 micro firm-level data from Nice Information Service from 1997 to 2009 to assess the economic contribution of financial systems. In general, a firm can choose financing methods from these four options: (1) the amount of bond issues, (2) the amount of paid-in-capital increases from the market-based financial system, (3) the amount of short-term loans, and (4) the amount of long-term loans from the bank-based financial system. To foster a firm's output growth, the local government uses monetary and fiscal policy tools, including credit guarantee volume, tax revenue, investment expenditures, and non-investment expenditures. These are classified by region (the Seoul metropolitan area and the rest of Korea[3]) and firm size (a large firm, a small- or medium-sized firm[4]).

Table 2 shows the shares of number of firms and of financing amount by financing option (e.g., bonds, paid-in-capital increases, short-term loans, and long-term loans). Firms in the Seoul metropolitan area (SMA) show a 5.2–6.2 % higher share of a number of firms issuing bonds compared to firms in the rest of Korea (ROK). However, firms in ROK record a larger share of financing amount by bonds issued until 2004 but a smaller share from 2005 to 2009. These figures imply that firms in ROK have more constraints on bond issues. The large firms tend to have more favorable terms for issuing bonds compared to small- and medium-sized firms in terms of management performance and internal reserve facilitation. Only a few firms are financed by paid-in-capital increases, and a large number of firms in

[3] The SMA includes Korea's capital city, Seoul, and its adjacent provinces, Incheon and Gyeonggi.

[4] In Korea, the definition of small- and medium-sized firms is specified in the Minor Enterprises Act.

Table 2 Financing options by region and firm (unit: %)

			SMA			ROK		
		Firm size	1997–2000	2001–2004	2005–2009	1997–2000	2001–2004	2005–2009
Bond	Share of number of firms	Total	19.5	16.8	17.4	13.3	11.2	12.2
		Large	43.2	35.0	32.6	33.9	26.7	25.1
		Small or medium	10.9	10.1	11.8	9.0	8.0	9.6
	Share of financing amount	Total	16.0	12.8	12.6	19.8	14.1	10.0
		Large	16.2	12.9	12.6	20.5	14.2	10.3
		Small or medium	13.4	9.8	11.8	15.5	11.5	7.3
Paid-in-capital increase	Share of number of firms	Total	0.0	1.3	3.9	0.0	0.7	2.6
		Large	0.0	2.5	5.6	0.0	2.1	3.9
		Small or medium	0.0	0.9	3.3	0.0	0.4	2.3
	Share of financing amount	Total	0.0	3.5	1.6	0.0	40.9	3.1
		Large	0.0	3.2	1.1	0.0	26.6	3.3
		Small or medium	0.0	5.0	10.8	0.0	7.9	4.8
Short-term loan	Share of number of firms	Total	81.4	74.0	72.7	85.4	79.4	79.2
		Large	82.7	70.6	64.9	80.4	69.5	67.3
		Small or medium	80.9	75.2	75.5	86.4	81.4	81.7
	Share of financing amount	Total	12.8	8.8	8.5	20.3	13.2	12.3
		Large	12.2	7.9	7.0	19.8	11.1	8.5
		Small or medium	18.1	15.9	19.4	21.1	17.5	20.0
Long-term loan	Share of number of firms	Total	72.0	58.6	52.8	81.8	71.8	65.6
		Large	78.4	63.8	55.7	79.6	66.2	53.2
		Small or medium	69.7	56.7	51.7	82.3	72.9	68.2
	Share of financing amount	Total	9.8	5.5	5.9	13.3	7.6	4.9
		Large	9.6	5.3	5.7	12.3	6.3	3.2
		Small or medium	11.4	7.9	8.8	15.7	10.7	9.4

Table 3 Monetary and fiscal policies of local government in Korea (unit: million US$)

	SMA			ROK		
	2000	2005	2009	2000	2005	2009
Average credit guarantee amounts per firm*	0.033	0.022	0.016	0.028	0.022	0.015
Local tax revenue	8482.9	15,789.4	19,776.2	5100.9	10,040.7	11,837.6
Investment expenditures of local governments	12,590.2	18,763.4	30,318.4	17,057.3	27,075.2	45,820.8
Non-investment expenditures of local governments	4211.2	10,239.2	13,202.0	4680.1	8244.5	10,594.1

*Data, which was supplied by the Small and Medium Business Administration, were aggregated annually

both regions are heavily dependent on short-term and long-term loans. In particular, firms in ROK tend to depend more on bank-based financing options (e.g., higher by 4.0–6.5 % in the share of a number of firms using short-term loans and by 9.8–12.8 % in the share of a number of firms using long-term loans). In addition, firms in ROK are more dependent on short-term loans than long-term loans.

Table 3 summarizes the monetary and fiscal policies of local governments in Korea. Total regional average credit guarantee amounts per firm have been continuously decreasing since 2000, and the average credit guarantee amounts in ROK are mostly lower than those in SMA. This implies that monetary constraints are placed on firms in ROK in terms of financial credit assistance of local governments. Moreover, after the 2008 financial crisis, local governments significantly reduced their credit offerings by 29.6–35.4 % in 2009 versus 2005.

3.2 *Analysis*

While most studies, including King and Levine (1993a, b) and Guiso et al. (2004), have argued that the development of financial systems affects the productivity and the output growth of firms, this paper examines the economic impacts of the financial sector (a bank-based or market-based financial system) and monetary and fiscal policies on a firm's growth. The basic form of the output growth model of firms, including the financial systems addressed in this paper, can be written as shown in Eq. (1):

$$\ln Y_{i,t} = a + \alpha \ln L_{i,t} + \beta \ln K_{i,t} + \sum_{k=1}^{4} \gamma_k \ln F_{i,t}^{k} + \varepsilon_{i,t} \tag{1}$$

where Y is output of firms, L labor, K capital, F financing options (bond issues, paid-in-capital increase, short-term loans, long-term loans), i firm, and t year.

The output growth of firms is expected to increase in response to the growth rate of the labor productivity, but this paper is more concerned with how the monetary resources of each financial system contribute to the output growth of firms, that is, the values of parameters, γ. This paper employs multilevel statistical models to analyze a hierarchical structure, *level 1* (firm level) and *level 2* (regional level).[5] The dependent variable is the total amount of sales of each firm (the output growth), and independent variables at the microlevel include (1) the total amount of tangible fixed assets and total number of employees as traditional input variables; (2) the amount of short-term and long-term loans, bond issues, and paid-in-capital increase as financing options; and (3) local credit guarantee amounts, local tax revenues, and investment and non-investment expenditures of local governments as regional monetary and fiscal policies. There are two dummy variables: financial crisis as economic shock and firm size. The final estimated equations of the two-level models can be written as follows and are estimated for 16 metropolitan regions from 1997 to 2009 in Korea.

3.2.1 Random Effect Model

$$\text{Level 1} \quad \ln y_{ij,t} = \beta_j^0 + \varepsilon_{ij} \qquad \varepsilon_{ij} \sim N(0, \sigma_\varepsilon^2) \tag{2}$$

$$\text{Level 2} \quad \beta_j^0 = \gamma^0 + v_j^0 \qquad v_j^0 \sim N(0, \sigma_v^2) \tag{3}$$

3.2.2 Random Intercept Model

$$\text{Level 1} \quad \ln y_{ij,t} = \beta_j^0 + \sum_{k=1}^{K} \beta_j^k \ln x_{ij,t}^x + \varepsilon_{ij} \qquad \varepsilon_{ij} \sim N(0, \sigma_\varepsilon^2) \tag{4}$$

$$\text{Level 2} \quad \beta_j^0 = \gamma^0 + \sum_{s=1}^{S} \gamma_j^{0s} \ln z_{j,t}^s + v_j^0 \qquad v_j^0 \sim N(0, \sigma_v^2) \tag{5}$$

3.2.3 Random Slope Model

$$\text{Level 1} \quad y_{ij,t} = \beta_j^0 + \sum_{k=1}^{K} \beta_j^k \ln x_{ij,t}^k + \varepsilon_{ij} \qquad \varepsilon_{ij} \sim N(0, \sigma_\varepsilon^2) \tag{6}$$

[5] In multilevel analysis, the hierarchical structure of data is explicitly considered (Goldstein 2003). Therefore, we could analyze the effects of regional monetary and fiscal policies among regions on the Level 2, using this model.

$$\text{Level 2} \quad \beta_j^k = \gamma_j^k + \sum_{s=1}^{S} \gamma_j^{ks} \ln z_{j,t}^s + \upsilon_j^k \qquad \upsilon_j^k \sim N(0, \sigma_\upsilon^2) \tag{7}$$

3.2.4 Random Intercept-Slope Model

$$\text{Level 1} \quad \ln y_{ij,t} = \beta_j^0 + \sum_{k=1}^{K} \beta_j^k \ln x_{ij,t}^k + \varepsilon_{ij} \qquad \varepsilon_{ij} \sim N(0, \sigma_\varepsilon^2) \tag{8}$$

$$\begin{aligned} \text{Level 2} \quad & \beta_j^0 = \gamma_j^0 + \upsilon_j^0 \\ & \beta_j^k = \gamma_j^k + \sum_{s=1}^{S} \gamma_j^{ks} \ln z_{j,t}^s + \upsilon_j^k \qquad \upsilon_j \sim N(0, \sigma_\upsilon^2) \end{aligned} \tag{9}$$

$y_{ij,t}$: output of firm i ($i = 1 \ldots m$) in region j ($j = 1 \ldots n$) and time t ($t = 1997 \ldots 2009$)
x: quantity variables including labor, capital, and financing options at firm level and quality (dummy) variables of financial crisis and firm size
z: regional level variables of monetary and fiscal policies

Table 4 shows the results of four multilevel statistical models and an ordinary least square (OLS) model. The model fitness of the random intercept-slope model is the highest among the four multilevel statistical models. The output inequality within a region is larger than that among regions in the sense that the intercept parameters of level 1 (0.741–2.361) are higher than those of level 2 (0.051–0.119) in the random effect (column 1–column 3). However, the output inequality among regions is widened in the random intercept-slope model because of the disparities of regional policies (see column 4). Importantly, the elasticity of the investment expenditures with respect to the output is estimated as 0.290, which is the highest among the regional macroeconomic variables. The local tax revenue and the non-investment expenditures of the local government positively influence the output of firms,[6] while the effect of credit guarantees of the local government on output of firms is insignificant. However, the fiscal policies could enlarge the disparity of regional economies.

With regard to financial systems (column 2–column 4), three financing options—short-term loans in the bank-based system, bonds, and paid-in-capital in the market-based system—have positive effects on the output of firms; if bond financing increased by 1 %, then the output would grow by 0.043 %. Only a long-term loan has a negative elasticity with respect to the output as much as (−) 0.003 to (−) 0.004 due to capital borrowing costs and the rollover burden of a long-term loan. This risk-averse activity of banks can reduce the economic risk from long-

[6] Niskanen (1971) explained that public administrators might be motivated to maximize revenue, and thus expenditures, in order to expand the desirable aspects of their working environment.

Table 4 Estimation of multilevel statistical models

Variable	Random effect	Random intercept	Random slope	Random intercept-slope	OLS
Fixed effect					
Intercept	10.311***	2.105***	0.320	0.115	10.524***
Firm-specific variables					
Log (labor)		0.721***	0.722***	0.722***	0.724***
Log (capital)		0.083***	0.082***	0.082***	0.076***
Firms' size dummy (large firms = 1)		0.403***	0.403***	0.403***	0.446***
Financial variables					
Log (short-term loan)		0.026***	0.026***	0.026***	0.028***
Log (long-term loan)		−0.004***	−0.003**	−0.004***	−0.002
Log (bond)		0.043***	0.043***	0.043***	0.041***
Log (paid-in-capital increase)		0.008*	0.007*	0.007*	−0.001
Regional macro variables					
Financial crisis dummy (1997–1999)		−0.530***	−0.285***	−0.239***	−0.662***
Log (credit guarantee)		−0.062***			
Log (local tax revenue)		0.017			
Log (investment expenditures)		0.553***			
Log (non-investment expenditures)		0.015			
Random effect					
Level 1					
Intercept	2.361***	0.743***	0.741***	0.740***	
Level 2					
Intercept	0.051***	0.119***		7.105***	
Regional macro variables					
Log (credit guarantee)			0.001*	0.001	
Log (local tax revenue)			0.025	0.083**	
Log (investment expenditures)			0.339***	0.290***	
Log (non-investment expenditures)			0.054*	0.014*	
Chi-square	686.95***	799.78***	1765.56***	1795.96***	
Adj R-Sq					67.62

*, **, and *** represent respectively the statistical significance at 10 %, 5 %, and 1 % levels

term financing with respect to profit maximization in the short run (Schmidt et al. 1999), but it negatively affects outputs of firms in the ROK, which are very dependent on short-term loans, as indicated in Table 2. The elasticity for output of firms is followed by bonds, short-term loans, and paid-in-capital increases. That is, the combination of bank financing for short-term periods (or for emergency purposes) and market financing for long-term periods (e.g., issuing bonds) could have positive effects on the output growth of firms. This means that the relationship between the market-based and the bank-based system could be complementary for the output growth of firms, as discussed in Deidda and Fattouh (2008). The economic contribution of large firms is higher than that of small- and medium-sized firms, but the output is negatively affected by the financial crisis shock (see column 2–column 4). Based on these results, we offer some observations about types of financial strategies for small- and medium-sized firms. First, credit rationing needs to be extended to small- and medium-sized firms in order to reduce lenders' cost of providing loans, such as information and monitoring costs, as discussed in Peterson and Rajan (1994). More credit with policy funds should be allocated to innovative small and medium firms based on a technology evaluation system in order to enhance the financial stability of firms with fewer collateral loans. As a long-term financing option, a private bond market focused on small- and medium-sized firms needs to be established to utilize investment funds of life insurance companies, universities, and pensions; the types of the bond will vary in terms of maturity, issue amount, options to redeem bonds, and coupon rate. In addition, a trading system of unlisted stocks and call and put options can contribute to diversifying the investors' options to recoup the investment amounts and promoting investments in paid-in-capital increases of small- and medium-sized and venture firms.

Table 5 presents the results of the output growth model by province in Korea. The effects of financing options on the output are similar to those shown in Table 4: short-term loans and bonds have positive influences on output growth, but long-term loans have a negative effect in most regions, while the effect of paid-in-capital increases is inconclusive and mostly insignificant. These parameters in Table 5 are regressed by monetary and fiscal variables, including local government credit guarantee, local tax revenue, investment expenditures, and non-investment expenditures of local government.

In Table 6, if there is a 1 % increase in the amount of credit guarantees from the local government, the elasticity values of factor inputs and financing options with respect to the output would increase by 0.794 % to 0.883 %. The impact is larger than the other cases: for example, a 1 % decrease in local tax rates would increase the elasticity values of factor inputs and financing options by 0.134 to 0.753 %. The investment expenditure of local governments positively affects the elasticity of paid-in-capital increase, while the effects on the other financing options are also positive but statistically insignificant. Thus, a credit guarantee as monetary policy

Table 5 Estimation of output growth model of firms for 16 metropolitan regions

Region	Seoul	Busan	Daegu	Incheon
Intercept	10.737***	10.688***	10.511***	10.457***
Log (labor)	0.898***	0.949***	0.965***	0.860***
Log (capital)	0.102***	0.051***	0.035*	0.140***
Firms' size dummy (large firms = 1)	0.062***	−0.041	0.015	0.473***
Log (short-term loan)	0.031***	0.024***	0.019**	0.017**
Log (long-term loan)	−0.003	−0.022***	−0.039***	−0.034***
Log (bond)	0.020***	0.037***	0.059***	0.011**
Log (paid-in-capital increase)	0.002	−0.005	−0.022	−0.002
Financial crisis dummy (1997–1999)	−0.654***	−0.713***	−0.682***	−0.685***
Adj R-Sq	68.98	50.85	63.90	70.72
Region	Gwangju	Daejeon	Ulsan	Gyeonggi
Intercept	10.801***	10.406***	10.647***	10.522***
Log (labor)	0.718***	0.824***	0.886***	0.837***
Log (capital)	0.282***	0.176***	0.114***	0.163***
Firms' size dummy (large firms = 1)	−0.066	0.177*	0.299***	0.162***
Log (short-term loan)	0.031**	0.029**	0.011	0.018***
Log (long-term loan)	−0.070***	−0.035***	0.009	−0.015***
Log (bond)	0.001	−0.008	0.004	0.019***
Log (paid-in-capital increase)	0.010	−0.099	−0.052	0.019*
Financial crisis dummy (1997–1999)	−1.013***	−0.057***	−0.862***	−0.625***
Adj R-Sq	49.92	57.77	77.11	67.45
Region	Kangwon	North Chungcheong	South Chungcheong	North Jeolla
Intercept	10.227***	10.488***	10.514***	10.661***
Log (labor)	0.883***	0.896***	0.893***	0.921***
Log (capital)	0.117***	0.104***	0.107***	0.079***
Firms' size dummy (large firms = 1)	−0.088	−0.157***	0.433***	−0.215**
Log (short-term loan)	0.001	0.020***	−0.006	0.026**
Log (long-term loan)	−0.003	−0.028***	−0.039***	−0.003
Log (bond)	0.033**	0.015*	0.041***	0.014
Log (paid-in-capital increase)	−0.087	0.004	−0.038*	0.023
Financial crisis dummy (1997–1999)	−0.642***	−0.634***	−0.608***	−0.619***
Adj R-Sq	57.33	54.33	63.60	64.43
Region	South Jeolla	North Gyeongsang	South Gyeongsang	Jeju
Intercept	10.560***	10.594***	10.613***	10.682***
Log (labor)	0.873***	0.856***	0.942***	0.750***
Log (capital)	0.127***	0.144***	0.058***	0.250***
Firms' size dummy (large firms = 1)	−0.126	0.043	0.086*	−0.951***
Log (short-term loan)	−0.015	−0.011*	0.023***	−0.158***

(continued)

Table 5 (continued)

Region	South Jeolla	North Gyeongsang	South Gyeongsang	Jeju
Log (long-term loan)	0.041***	0.002	−0.020***	0.076***
Log (bond)	−0.004	0.029***	0.024***	0.098
Log (paid-in-capital increase)	−0.020	0.007	0.011	−1.164
Financial crisis dummy (1997–1999)	−0.772***	−0.675***	−0.722***	−0.432*
Adj R-Sq	65.70	65.88	63.42	48.05

*, **, and *** represent respectively the statistical significance at 10 %, 5 %, and 1 % levels

Table 6 Estimation of each parameter in output growth model of firms

Dependent variable	Intercept	Log (credit guarantee)	Log (local tax revenue)	Log (non-investment expenditures)	Log (investment expenditures)
Intercept	5.882***	0.798***	−0.643*	0.658	0.187
Labor	−3.338*	0.840***	−0.753*	0.696	0.208
Capital	−3.675**	0.883***	−0.582*	0.481	0.219
Short-term loan	−4.246***	0.848***	−0.501*	0.394	0.259
Long-term loan	−4.286**	0.845***	−0.618*	0.497	0.275
Bond	−4.128**	0.853***	−0.616*	0.524	0.239
Paid-in-capital increase	−5.229***	0.794***	−0.134	−0.109	0.449**

*, **, and *** represent respectively the statistical significance at 10 %, 5 %, and 1 % levels

and a reduction in the local tax rate as fiscal policy are key policies to increase the output of firms and to promote financing in markets. Consequently, local governments should extend their economic role from simply offering credit for loans to directly buying firms' bonds or, if necessary, taking over paid-in-capital increases, because of the larger effect of monetary policy. In addition, their focus should be on an allocation of fiscal expenditures and tax reductions to improve the efficiency of fiscal policies. In terms of complementary relations between the two financial systems, firms can implement the best financing option to minimize capital costs and financial risk under financial stability. That is, the government and financial authorities need to develop a crowding fund, a hedge fund, and a FinTech in the market-based system by mitigating regulations. Also, the financial authority needs to encourage banks to invest start-ups and venture companies aggressively through easing excessive prudential regulations.

4 Conclusion and Policy Implications

This paper explores how the output growth of firms depends on monetary and fiscal policies. It finds that short-term loans and bonds have positive influences on output growth, but long-term loans have a negative effect in most regions. A credit guarantee from a local government has the largest impact on the elasticity values of factor inputs and financing amounts with respect to output compared to any other government expenditure program. Also, a credit guarantee from a local government and a reduction in the regional tax burden are key policies to increase the output of firms and to promote financing in markets.

One future research area to examine is the efficiency of the legal system, including financial inspection of the terms enforcing investors' rights, with the aim of improving the long-run growth and stability of financial systems. The efficiency could be examined through data envelop analysis or a stochastic frontier analysis method. Another issue to explore is the casual relation between financial systems and economic growth on the basis of regional level. This would account for regional differences in long-run regional economic performance due to changes in financial structure. Finally, more attention should be paid to analyzing the impacts of financial systems and financing options on the business cycle of industry sectors and markets. Because the magnitudes and signs of the impacts are expected to rely on the phases of industrial growth and life cycle stages of the market, this research may identify suitable financial packages for suppliers.

References

Beare, J.B. 1976. A monetarist model of regional business cycles. *Journal of Regional Science* 16: 57–63.

Beck, T., and R. Levine, 2001. *Stock markets, banks, and growth*, Policy research working paper, 2670. The World Bank.

Becketti, S., and C. Morris. 1992. Does money still forecast economic activity? Federal reserve bank of Kansas city. *Economic Review* 77: 65–77.

Boocock, G., and M.N.M. Shariff. 2005. Measuring the effectiveness of credit guarantee schemes: Evidence from Malaysia. *International Small Business Journal* 23: 427–454.

Boyd, J.H., and M. Gertler. 1994. The role of large banks in the recent U.S. Banking crisis. *Federal Reserve Bank of Minneapolis Quarterly Review* 18(1): 1–24.

Chen, A., and N. Groenewold. 2011. Regional equality and national development in china: Is there a trade-off? *Growth and Change* 42(4): 628–669.

Cowling, M., and P. Mitchell. 1997. The evolution of U.K. self-employment: A study of government policy and the role of the macro economy. *The Manchester School* 65: 427–442.

Cumming, D. 2007. Government policy towards entrepreneurial finance: Innovation investment funds. *Journal of Business Venturing* 22(2): 193–235.

Deidda, L., and B. Fattouh. 2008. Banks, financial markets and growth. *Journal of Financial Intermediation* 17(1): 6–36.

Franks, J., and C. Mayer. 1993. *German capital markets, corporate control and the obstacles to hostile takeover: Lessons from three case studies*, Working paper. London Business School.

Gerschenkron, A. 1962. *Economic backwardness in historical perspective*. Cambridge, MA: Harvard.
Goldstein, H. 2003. *Multilevel statistical models*. London: Arnold.
Guiso, L., T. Jappelli, M. Padula, and M. Pagano. 2004. Financial market integration and economic growth in the EU. *Economic Policy* 19(40): 524–577.
Harrison, R.T., C.M. Mason, and P. Girling. 2004. Financial bootstrapping and venture development in the software industry. *Entrepreneurship and Regional Development* 16(4): 307–333.
Hooks, L., and T.C. Opler. 1993. The determinants of corporate bank borrowing. *Federal Reserve Bank of Dallas Financial Industry Studies*: 1–93.
Jacobides, M.J., T. Knudsen, and M. Augier. 2006. Benefiting from innovation: Value creation, value appropriation and the role of industry architectures. *Research Policy* 35(8): 1200–1216.
King, R.G., and R. Levine. 1993a. Finance, entrepreneurship and growth: Theory and evidence. *Journal of Monetary Economics* 32(3): 513–542.
King, R.G., and R. Levine. 1993b. Finance and growth: Schumpeter might be right. *The Quarterly Journal of Economics* 108(3): 717–737.
Kreft, S.F., and R. Sobel. 2005. Public policy, entrepreneurship, and economic freedom. *Cato Journal* 25(3): 595–616.
Lee, B.S. 2012. Bank-based and market-based financial systems: Time-series evidence. *Pacific-Basin Finance Journal* 20: 173–197.
Lerner, J. 2002. When bureaucrats meet entrepreneurs: The design of effective 'Public Venture Capital' programs. *The Economic Journal* 112: 73–84.
Levine, R. 2002. Bank-based or market-based financial systems: Which is better? *Journal of Financial Intermediation* 11: 398–428.
Levine, R., and S. Zervos. 1998. Stock markets, banks, and economic growth. *The American Economic Review* 88(3): 537–558.
Li, W. 2002. Entrepreneurship and government subsidies: A general equilibrium analysis. *Journal of Economic Dynamics and Control* 26(11): 1815–1844.
McGreal, S., J. Berry, G. Lloyd, and J. McGarthy. 2002. Tax-based mechanisms in urban regeneration: Dublin and Chicago models. *Urban Studies* 39(10): 1819–1831.
Meager, N., P. Bates, and M. Cowling. 2003. An evaluation of business start-up support for young people. *National Institute Economic Review* 186: 59–72.
Niskanen, W. 1971. *Bureaucracy and representative government*. Chicago: Aldine-Atherton.
Oh, I., J. Lee, A. Heshmati, and G. Choi. 2009. Evaluation of credit guarantee policy using propensity score matching. *Small and Business Economics* 33: 335–351.
Opler, T.C. 1993. Controlling financed distress costs in leveraged buyouts with financial innovations. *Financial Management* 22(3): 79–90.
Peterson, M.A., and R.G. Rajan. 1994. The benefits of lending relationships: Evidence from small business data. *Journal of Finance* 49(1): 3–37.
Porter, M.E. 1992. Capital choices: Changing the way America invests in industry. *Journal of Applied Corporate Finance* 5(2): 4–16.
Rajan, R.G., and L. Zingales. 1998a. Financial dependence and growth. *The American Economic Review* 88(3): 559–586.
Rajan, R.G., and L. Zingales. 1998b. Which capitalism? Lessons from the East Asian crisis. *Journal of Applied Corporate Finance* 11(3): 40–48.
Rioja, F., and N. Valey. 2004. Does one size fit all? A reexamination of the finance and growth relationship. *Journal of Development Economics* 74: 429–447.
Rousseau, P.L., and P. Wachtel. 2002. Inflation thresholds and the finance-growth nexus. *Journal of International Money and Finance* 21(6): 777–793.
Schmidt, R.H., A. Hackethal, and M. Tyrell. 1999. Disintermediation and the role of banks in Europe: An international comparison. *Journal of Financial Intermediation* 8: 36–67.
Shin, D.J., and B.H.S. Kim. 2011. Efficiency of the banking industry structure in Korea. *Asian Economic Journal* 25(4): 355–373.

Stulz, R.M. 2000. Financial structure, corporate finance, and economic growth. *International Review of Finance* 1(1): 11–38.

Takii, K. 2008. Fiscal policy and entrepreneurship. *Journal of Economic Behavior and Organization* 65(3–4): 592–608.

Yu, Y., and D.S. Rickman. 2013. US state and local fiscal policies and non-metropolitan area economic performance: A spatial equilibrium analysis. *Papers in Regional Science* 92(3): 579–598.

Part III
Disaster and Impact Analysis

Assessment of Community Vulnerability to Natural Disasters in Korea by Using GIS and Machine Learning Techniques

Dong Keun Yoon and Seunghoo Jeong

Abstract Despite similar natural hazard magnitudes, the economic losses and fatalities due to natural disasters are usually unevenly distributed among nations, regions, communities, and individuals. Socially, economically, and environmentally vulnerable communities are more likely to suffer disproportionately from disasters. Identifying vulnerability factors to disasters is critical information for disaster managers and planners to make disaster-related policy and strategies for mitigating the negative impacts of disasters. This study constructs an index of disaster vulnerability of local communities in Korea. Twelve indicators including social, economic, and natural environment and built environment aspects are selected to assess 230 local communities' vulnerability to disasters. Economic losses from disasters from 2001 to 2010 in Korea are analyzed using GIS. Moreover, this study examines the relationships between the constructed vulnerability indicators and economic damage from natural disasters. Machine learning techniques including Cubist and Random Forest are applied to examine what vulnerability indicators are statistically associated with disaster damage in Korea.

Keywords Community vulnerability • Disaster damages • GIS • Machine learning

1 Introduction

Despite similar natural hazard magnitudes, natural disasters result in unevenly distributed damage within or among nations, regions, communities, and individuals (Yoon 2012). Socially, economically, and environmentally vulnerable countries, regions, communities, and individuals are more likely to suffer disproportionately

D.K. Yoon (✉)
Department of Urban Planning and Engineering, Yonsei University, Seoul, Korea
e-mail: dkyoon@yonsei.ac.kr

S. Jeong
School of Urban and Environmental Engineering, Ulsan National Institute of Science and Technology, Ulsan, South Korea
e-mail: shjeong@unist.ac.kr

E. Kim, B.H.S. Kim (eds.), *Quantitative Regional Economic and Environmental Analysis for Sustainability in Korea*, New Frontiers in Regional Science: Asian Perspectives 25, DOI 10.1007/978-981-10-0300-4_7

from disasters. Identifying characteristics of vulnerability to natural disasters provides critical information to emergency managers and planners for making disaster-related policy and strategies to mitigate the adverse impacts of disasters (Emrich and Cutter 2011; Yoon 2012).

One of the significant issues in vulnerability assessment is to select vulnerability indicators. Two methods, the inductive and deductive approaches, are widely accepted to select vulnerability indices in vulnerability studies (Yoon 2012). Whereas the deductive approach focuses on theoretical relationships, the inductive approach is based on the statistical relationship between vulnerability indicators. Ordinary least squares (OLS) regression has usually been used to examine the vulnerability indicators regardless of the methodology for indicator selection (Cutter et al. 2003; Yoon 2012). OLS regression, however, is likely to show weak model performance because of data assumptions, including no correlation between model residuals and constant variance over space (Hamilton 1992; Tu and Xia 2008).

Recently, researchers have shown an increased interest in machine learning approaches to regression analysis in the fields of remote sensing, housing policy, and water environment (Grömping 2009; Kim et al. 2014; Li et al. 2014; Yoo et al. 2012). Unlike traditional OLS regression with data assumptions, machine learning approaches have shown better model performance in that no data assumptions are required (Kim et al. 2014). Despite the advantages of machine learning methods, there have been few studies in the vulnerability literature using machine learning approaches.

The aim of this study is to construct an index of disaster vulnerability for local communities in Korea. This study first gives a brief overview of the importance of vulnerability assessment and methodologies for variable selection. Based on the deductive approach, 12 indicators including social, economic, and natural environment and built environment aspects are selected to assess 230 local communities' vulnerability to natural disasters. Economic damages from natural disasters are analyzed using GIS from 2001 to 2010. Moreover, this study examines the relationship between selected vulnerability indicators and economic damage from natural disasters using OLS regression. Machine learning techniques are applied to examine what vulnerability indicators are associated with disaster damage. The results from two machine learning techniques, Random Forest and Cubist, are compared with those of traditional OLS regression to examine the significant vulnerability factors in Korea. Further, the model performances of each regression technique are compared to examine what regression method is more accurate for predicting disaster damage.

2 Community Vulnerability Assessment

There is a large volume of published studies placing emphasis on the importance of vulnerability assessment to natural disasters (Uitto 1998; Cutter and Finch 2008; Emrich and Cutter 2011; Yoon 2012; Gao et al. 2014). Assessing vulnerability to natural disasters is helpful for the development of emergency plans for disaster

(Uitto 1998), the improvement of emergency management (Cutter and Finch 2008; Gao et al. 2014), the development and implementation of mitigation strategies for damage reduction (Emrich and Cutter 2011), and the identification of causes of the different levels of damage (Yoon 2012).

Among numerous studies identifying vulnerability to natural disasters, the Social Vulnerability Index (SoVI) suggested by Cutter et al. (2003) has been widely used in various kinds of vulnerability assessment (Emrich and Cutter 2011; Yoon 2012; Zhou et al. 2014). Cutter et al. (2003) suggested SoVI as an index of vulnerability to natural disaster to find the relative vulnerability of counties in the United States. By using the inductive approach for variable selection, SoVI was constructed from 42 variables associated with social, economic, natural, and built environmental aspects.

The concept of SoVI made a contribution to the broader field of vulnerability studies. The SoVI concept was used to identify the effect of social vulnerability on migration patterns after hurricanes Katrina and Rita (Myers et al. 2008). The study by Emrich and Cutter (2011) examined the spatially distributed social vulnerability to drought, flooding, hurricane winds, and sea level rise in the Southern United States by using the concept of SoVI. Vulnerability to earthquakes was examined not only by using SoVI but also by using spatial multi-criteria social vulnerability (SEVI model) based on a weighting scheme in Bucharest, Romania (Armaş and Gavriş 2013).

3 Methods

3.1 Study Area

Korea as a study area consists of seven metropolitan cities (Seoul, Busan, Daegu, Incheon, Gwangju, Daejeon, Ulsan) and nine provinces (Gyeonggi, Gangwon, North Chungcheong, South Chungcheong, North Jeolla, South Jeolla, North Gyeongsang, South Gyeongsang, Jeju). As of 2010, there are 230 local communities belonging to these metropolitan cities and provinces (Korean Statistical Information Service 2011). These local communities are classified as "si," "gu," or "gun." While a province consists of cities (si) or counties (gun) which are legal subdivisions of the province, each metropolitan city is divided into districts (gu). There are 74 "gu" (district) within the seven metropolitan cities and 156 "si" (city) and "gun" (county) within the nine provinces.

3.2 Dependent Variable

By using machine learning techniques (Random Forest, Cubist) as well as OLS regression, this study examined what kinds of vulnerability indicators are related to economic damage. For this analysis, the annual average economic loss due to natural disasters during a 10-year period (2001–2010) was used as the dependent variable. The economic damages from 2001 to 2010 were normalized through the inflation rate in 2010 (the inflation rate in 2010 = 1). The dependent variable in 230 local communities was log-transformed to improve its distribution.

Typhoon, heavy rain, heavy snow, and snowstorm have been the main types of natural disaster influencing on Korea in terms of human and economic losses. Over 10 years (2001–2010), the annual economic losses due to natural disasters were as high as 1903 million US dollars in Korea (Fig. 1), and 135 human casualties (death, injured, and missing) occurred annually. Whereas typhoon, snowstorm, and flood were the main causes of economic losses, human casualties were inclined to occur from typhoon and flood. During this period, there were two catastrophic typhoons named “Rusa” (2002) and “Maemi” (2003) which are the first and second most severe typhoons in the history of Korea causing 7000 million and 5000 million US dollars of economic losses, respectively. The economic losses, however, have a tendency to decrease from 1500 million US dollars in 2001 to 390 million US dollars in 2010.

The per capita economic losses caused by natural disasters are unevenly spatially distributed within Korea (Fig. 2). From 2001 to 2010, the economic losses were concentrated in the northeast (Gangwon) and south-central (North Jeolla, South Gyeongsang, North Gyeongsang) parts of Korea, amounting to more than 3500 US dollars per capita. Although about 24 million people (49.8 % of the total population Korea in 2010) live in the Seoul Metropolitan Area (Seoul, Gyeonggi, Incheon), there were less economic losses per capita, 10 US dollars in Seoul, 60 US dollars in Incheon, and 170 US dollars in Gyeonggi region compared with Gangwon region (4500 US dollars) (Fig. 2).

3.3 Independent Variables

This study deductively selected 12 vulnerability indicators related to people (demographic and economic aspects) and places (natural and built environment aspects) based on previous studies (Cutter et al. 2003; Zahran et al. 2008; Yoon 2012). A z-score transformation was used to standardize the selected indicators by removing the different unit of each indicator. A detailed explanation of selected indicators is provided in Table 1.

This study measured and analyzed three vulnerability indicators to represent demographic aspects. The first demographic indicator is less educated people, measured by the percent of the population over 15 years of age who have not

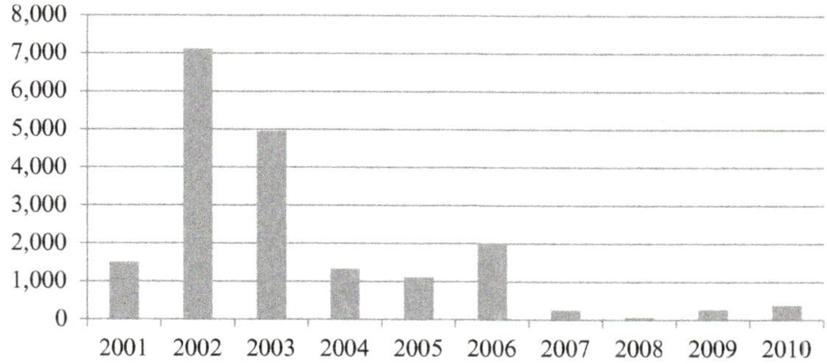

Fig. 1 Normalized damage from natural disasters in Korea (million USD, 2001–2010)

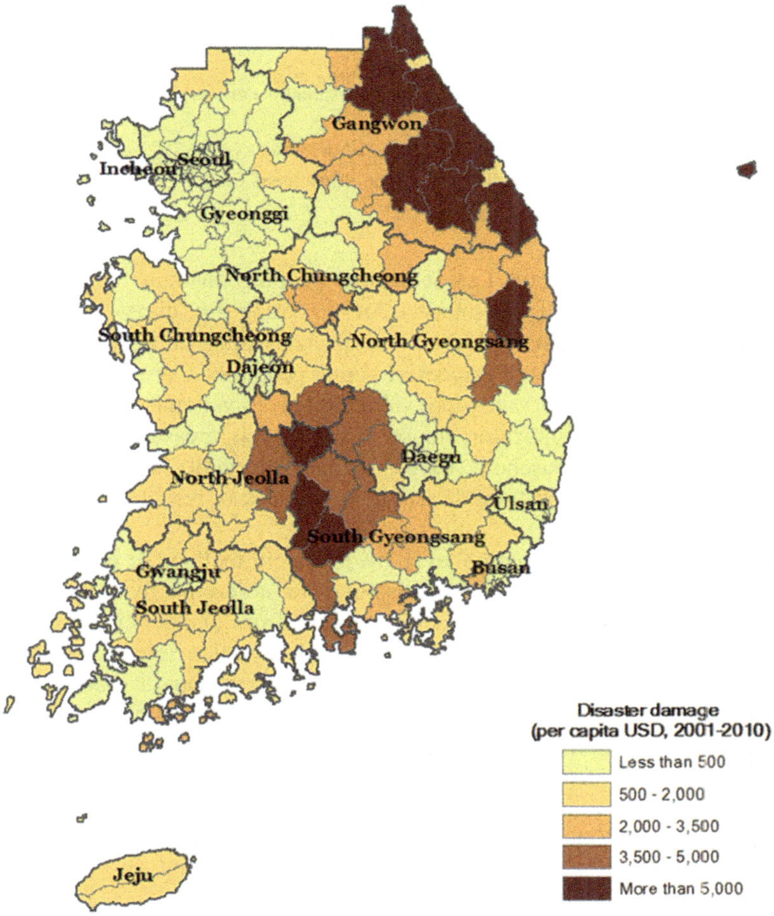

Fig. 2 Spatial distribution of disaster damage (2001–2010)

Table 1 Descriptions of indicators and expected sign for disaster damages

Indicators	Description	Year	Effect
Demographic aspects			
Less educated people	Percent of population over 15 years of age without high school completion	2010	+
Young people	Percent of population under 15 years of age	2010	+
Household size	Average number of people per household	2010	+/−
Economic aspects			
Greenhouses	Percent area of greenhouses for farming	2010	+
Tax	Per capita property tax	2010	−
Small business	Percent of small businesses with less than 30 employees	2010	+
Natural environment aspects			
Open space	Percentage of open space	2010	−
Precipitation	Annual precipitation between 2001 and 2010	2010	+
Low-lying area	Low-lying area below 10 m elevation above sea level	2010	+
Built environment aspects			
Industry density	Number of manufacturing establishments per square kilometer	2010	+/−
Old building	Percent of constructions built before 1959	2005	+
Urban population	Percent of population living in urban areas	2010	+/−

graduated from high school. Less educated people constrain the capability to understand and access to information for preparing for disasters in advance, increasing the vulnerability to natural disasters (Cutter et al. 2003; McEntire 2012). This study selected the percent of young people who are under 15 years of age as a second indicator showing vulnerability to natural disasters (Martins et al. 2012). The third indicator, household size, was measured as the average number of people per household. Large families, especially with the elderly, are likely to be vulnerable to natural disaster due to a lack of resources for supporting families (Cutter et al. 2003).

Individuals with a low income level are relatively inclined to be vulnerable to natural disasters due to their lack of resources to store emergency supplies, buy flood insurance, and have the capacity to recover from disasters (McEntire 2012; Felsenstein and Lichter 2014). Communities with a weak economic condition are also vulnerable to natural disasters due to a similar rationale as poor individuals (McEntire 2012; Felsenstein and Lichter 2014). This study estimated and modeled three vulnerability indicators representing economic conditions for communities and individuals. The first indicator was greenhouses, measured by the percent area of greenhouses for farming in a community. Previous studies have indicated that workers in primary industries are likely to be severely affected by natural disasters (Zhou et al. 2014; Chang et al. 2015). Second, this study measured the amount of property tax per total number of population in a community. As an index of income level, an individual or household with less property assets is likely to have low level of income, and this makes them vulnerable to natural disaster (Esnard et al. 2011;

Kuhlicke et al. 2011). Property tax was utilized to represent housing property from the study of flood risk assessment in Korea (Kim et al. 2012). The third economic indicator calculated the number of small businesses, defined as those with less than 30 employees, per total number of businesses in a community. Small businesses are likely to be vulnerable to disaster because of a lack of disaster planning, interruption of cash flow, and lack of capital for recovery (Runyan 2006).

This study selected three vulnerability predictors related to natural environment aspects: open space, precipitation, and low-lying area. To measure the open space, this study calculated the percentage of park area in a community. Parks include city parks, neighborhood parks, children's parks, cemetery parks, sports parks, waterside parks, culture parks, and historical parks. The higher percentage of open space a community has, the more resilient it becomes to natural disasters (Schipper and Pelling 2006; Biass et al. 2013). The second indicator, precipitation, was measured as the annual amount of precipitation from 2001 to 2010, which is one of the factors that increase vulnerability to floods (Fengqing et al. 2005; Zahran et al. 2008). Low-lying areas below 10 m elevation above sea level, as another environmental indicator, are vulnerable to flooding and storm damage.

The built environment aspects include three vulnerability indicators related to disaster damage. This study used the industry density which is the number of manufacturing establishments per square kilometer. Contrary to a greenhouse, manufacturing establishments are more likely to be located in urban areas. Usage of both greenhouses and industry density compares rural areas with greenhouses and urban regions with manufacturing establishments to analyze the communities' economic condition. Previous studies have considered the density of manufacturing establishments as a potential factor affecting the level of vulnerability in a community (Flax et al. 2002; Burton and Cutter 2008). The second indicator, old building, was measured as the percentage of housing units built before 1959. As old buildings are physically vulnerable to disaster due to the superannuation (Yoon 2012), a community with a high portion of old buildings is likely to have disaster damage (Zhou et al. 2014). The last indicator, urban population, was calculated as the percent population living in urban areas. Increased population growth and density increase human exposure and vulnerability to disasters (Perrow 2011). The degree of human and economic losses is associated with the population and infrastructure density of the area affected from disasters.

This study analyzed the correlation matrix among 12 selected independent variables to identify whether there is redundancy among indicators or not (Table 2). Checking the redundancy is important to obtain a parsimonious regression model. The correlation matrix shows that multicollinearity is not an issue since there are no correlation coefficients greater than 0.80, which is generally accepted cutoff value for correlations among variables (Hair et al. 1998; Malhotra et al. 1999).

Table 2 Correlation between independent variables

	A	*B*	*C*	*D*	*E*	*F*	*G*	*H*	*I*	*J*	*K*
B	.252†										
C	−.529†	.443†									
D	.194†	.097	−.019								
E	−.398†	.086	.326†	−.047							
F	.313†	.212†	−.028	−.206†	−.358†						
G	−.491†	−.147*	.269†	−.242†	.163*	−.104					
H	−.149*	−.031	.008	−.025	−.103	−.108	.037				
I	.165*	.234†	−.084	−.017	.060	.025	−.108	.092			
J	−.275†	−.307†	−.028	−.307†	.029	−.075	.090	.038	−.129		
K	.654†	.433†	−.293†	.085	−.246†	.318†	−.292†	−.043	.140*	−.252†	
L	−.540†	−.402†	.135*	−.408†	−.009	−.029	.387†	.105	−.189†	.649†	−.413†

A less educated people, *B* young people, *C* household size, *D* greenhouses, *E* tax, *F* small business, *G* open space, *H* precipitation, *I* low-lying area, *J* industry density, *K* old building, *L* urban population

$*p < 0.05$, $†p < 0.01$

3.4 Machine Learning Techniques: Random Forest and Cubist

Introduced by Breiman (2001), Random Forest has become popular and widely used to implement classification and regression analysis. This technique has attracted a lot of attention in many diverse fields, such as biostatistics, the housing market, and remote sensing (Grömping 2009; Yoo et al. 2012; Kim et al. 2014).

As an ensemble of trees, a Random Forest has two different levels of randomness for improving the robustness of the training model. The first level of randomness is that each tree is generated by the randomly selected subset of training data. Second, at each node, a randomly selected subset of independent variables is used to determine a tree split (Breiman 2001; Grömping 2009). Random Forest models are composed of a series of binary rule-based decisions determining the extent of the relationship between the input data and an individual variable. Whereas all predictors are used to split nodes in classical classification and regression, Random Forest splits nodes using a random subset of independent variables at each node (Breiman 2001; Liaw and Wiener 2002).

This study implemented the Random Forest model by using the R statistical software (version 3.1.0; http://www.r-project.org/). As noncommercial software, the R statistical software provides the *randomForest* add-on package based on the study (Breiman 2001). This study used the default values provided by the *randomForest* add-on package, except for the default number of trees. One thousand trees were used instead. This study selected 161 (70 %) out of 230 local communities that were used in OLS regression for a training data set, and then the training model was constructed by Random Forest. The remaining data set (30 %) was used to validate the training model.

As commercially available software for data mining, Cubist is a powerful package for generating rule-based linear regression models in the form of multivariate regression (Yoo et al. 2012; Kim et al. 2014). Unlike Random Forest, which generates a single constant value at each final node, Cubist produces rule-based linear regression output based on instance-based criteria (Kim et al. 2014). Developed by RuleQuest Research Inc., this software has been widely used in a variety of studies related to classification and regression models. Cubist has the merit of analyzing substantial data sets containing numerous records quickly compared to neural networks or similar modeling. The classification and regression algorithms, however, are unknown because this software is not open to the public.

As Cubist does not require any data assumptions, Cubist models give better results than those produced by simple techniques such as OLS regression while also being easier to understand than neural networks. The rule-based regression output generated by Cubist makes regression results more uncomplicated and interpretable than Random Forest (Kim et al. 2014). Due to the advantages of Cubist, this regression technique has been widely used in the estimation of forest biomass (Im et al. 2009), the prediction of soil properties (Minasny and McBratney 2008), the quantification of impervious surface (Im et al. 2012), the estimation of house values using satellite images (Yu and Wu 2006), and the selection of hedonic variable (Yoo et al. 2012).

This study constructed a rule-based linear regression model to predict the economic losses due to natural disasters by using four explanatory factors, including demographic aspects, economic aspects, natural environment aspects, and built environment aspects. The training model was constructed by using a training data set (70 %) of 230 local communities used in the OLS regression model, and the remaining data set (30 %) was used to validate the training model.

4 Results

4.1 The Assessment of Community Vulnerability

Three regression techniques were used to examine what kinds of factors are associated with community vulnerability to natural disasters. This study compared the vulnerability indicators examined by three different regression models. For the comparison with the results from machine learning approaches, this study conducted an OLS linear regression model to identify the community vulnerability to natural disasters using the data from 161 randomly selected local communities (70 %) for the training model. The data from the remaining communities (30 %) were used to validate the training model. From the results of the OLS regression, five indicators, less educated people, greenhouses, precipitation, industry density, and urban population, were selected as the significant vulnerability indicators with a 95 % or 99 % statistical significance level (Table 3).

The result of this study indicates that the level of vulnerability to natural disasters is especially higher in a community with a high portion of less educated people, a high percentage of greenhouse area, high annual precipitation, a low density of industry, and a low portion of the population living in the urban regions.

The figure below illustrates the results obtained from the regression analysis of vulnerability indicators by Random Forest using training data (Fig. 3). The value of the increase in mean-square-error (IncMSE) in percentage is examined to indicate the relative importance of each. Urban population was determined to be the most critical vulnerability indicator, whereas low-lying area was the least important vulnerability indicator. Based on the study by Díaz-Uriarte and De Andres (2006), this study selected 80 % of the total indicators with the largest importance as significant indicators, including urban population, industry density, less educated people, greenhouses, old building, precipitation, open space, household size, and tax.

The regression model by Cubist provides a rule-based linear equation with vulnerability indicator usage information. The relative importance is determined by the coefficient value of each vulnerability indicator in a rule-based linear model. The three rule-based linear equations were obtained by the values of greenhouses

Table 3 Relationship between disaster damage and vulnerable population and place

Variables	B	t
Demographic aspects		
Less educated people	0.142*	2.054
Young people	0.124	1.970
Household size	−0.118	−1.774
Economic aspects		
Greenhouses	0.148**	3.477
Tax	−0.048	−0.991
Small business	−0.071	1.584
Natural environment aspects		
Open space	−0.071	−1.767
Precipitation	0.305**	2.635
Low-lying area	−0.048	−1.327
Built environment aspects		
Industry density	−0.447**	−5.757
Old building	−0.049	−0.954
Urban population	−0.133*	−2.070
Constant	5.491	
Adjusted R-square	0.659	
Number	161	
Durbin-Watson	1.941	
RMSE	0.291	
Significance	.000	

*$p < 0.05$, **$p < 0.01$

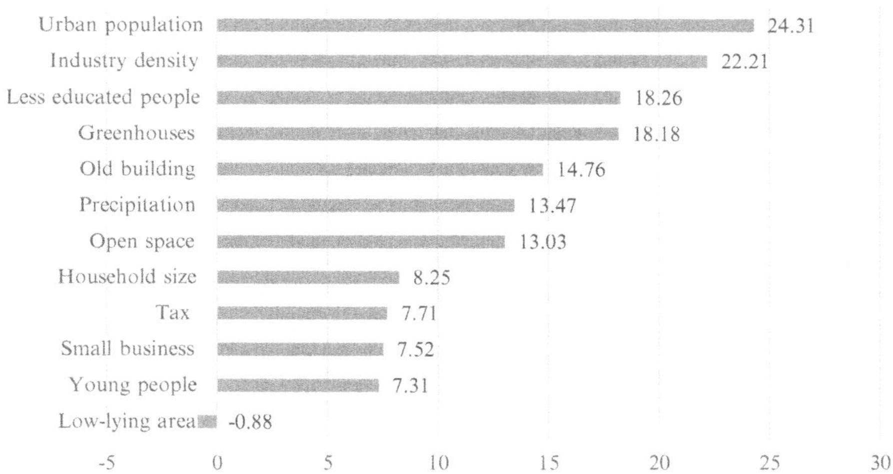

Fig. 3 % IncMSE from Random Forest

and urban population. The first rule is the dominant rule used in 77 of 161 communities when the value of tax is less than or equal to 1.87, the value of industry density is less than or equal to −0.3, and the value of old building is less than or equal to 0.89, respectively. Ten variables, industry density, urban population, precipitation, tax, household size, old building, low-lying area, small business, greenhouses, and less educated people, were selected as important vulnerability variables.

Rule 1 (77 samples): If $D \leq 1.87, A \leq -0.3, F \leq 0.89$,

$$\begin{aligned}\text{Damage}_{2001-2010} = \; & 4.212 - 4.05A - 0.641B + 0.202D - 0.113E + 0.103F \\ & + 0.32C - 0.05G + 0.045H + 0.026I + 0.017J\end{aligned}$$

Rule 2 (58 samples): If $A > -0.3$,

$$\begin{aligned}\text{Damage}_{2001-2010} = \; & 5.835 + 0.337I - 0.36A - 0.106H + 0.057J - 0.046B \\ & + 0.1C\end{aligned}$$

Rule 3 (3 samples): If $D > 1.87$,

$$\begin{aligned}\text{Damage}_{2001-2010} = \; & 4.866 + 0.153JA + 0.3C + 0.08\text{I} - 0.1A - 0.052D \\ & - 0.044B - 0.036F - 0.035G\end{aligned}$$

Rule 4 (23 samples): If $F > 0.89$,

$$\begin{aligned}\text{Damage}_{2001-2010} = \; & 5.153 - 3.27A + 0.073D + 0.11C - 0.037E + 0.02I \\ & - 0.018G\end{aligned}$$

where *A, B, C, D, E, F, G, H, I,* and *J* are industry density, urban population, precipitation, tax, household size, old building, low-lying area, small business, greenhouses, and less educated people, respectively.

The most striking result to emerge from the three regression results was that three different regression models similarly identified the significant vulnerability indicators related to disaster damage (Table 4). Critical vulnerability indicators examined by OLS regression, Random Forest, and Cubist had five vulnerability indicators in common out of 12: industry density, precipitation, greenhouses, less educated people, and urban population. The rank of importance of each indicator, however, was different for each regression technique. Industry density was the most significant variable in the results from OLS regression and Cubist and second in the Random Forest. Although the urban population indicator was the most important in the Random Forest and the second most important in Cubist, this indicator showed the fifth significance in OLS regression.

Table 4 Importance of vulnerability indicators depending on regression techniques

Rank	OLS regression	Random Forest	Cubist Rule 1
1	Industry density[a]	Urban population[a]	Industry density[a]
2	Precipitation[a]	Industry density[a]	Urban population[a]
3	Greenhouses[a]	Less educated people[a]	Precipitation[a]
4	Less educated people[a]	Greenhouses[a]	Tax[a]
5	Urban population[a]	Old building[a]	Household size[a]
6	Young people	Precipitation[a]	Old building[a]
7	Household size	Open space[a]	Low-lying area[a]
8	Small business	Household size[a]	Small business[a]
9	Open space	Tax[a]	Greenhouses[a]
10	Old building	Small business	Less educated people[a]
11	Low-lying area	Young people	
12	Tax	Low-lying area	

[a]Selected vulnerability indicators by three different regression techniques

4.2 Model Performance

The training and validation scatter plots between the observed economic losses and the predicted values by each regression model are illustrated in Fig. 4. The results of training data scatter plots indicated that the Random Forest regression model showed the best model fit with R^2 of 0.9458, which means this model can explain 94.58 % of the total variance. The Cubist regression model showed the second highest model fit with R^2 of 0.7583, which explains 75.83 % of the total variance. The OLS regression model, however, produced the lowest model fit with R^2 of 0.6585, which explains 65.85 % of the total variance. The conclusion is that machine learning techniques, both Random Forest and Cubist, showed higher model performance than the OLS regression model.

The results of the validation data scatter plots indicate that the Cubist regression model showed the best model fit with R^2 of 0.6346, which means this model can explain 63.46 % of the total variance. The Random Forest regression model showed the second highest model fit with R^2 of 0.6201, which explains 62.01 % of the total variance. The OLS regression model, however, indicated the lowest model fit with R^2 of 0.4954, which explains 49.54 % of the total variance. Similar to the training data, the model performance of machine learning techniques was higher than that of OLS regression.

Each regression technique provides its index for accuracy measurement. For the consistent comparison of model accuracy, root-mean-square-error (RMSE) and relative RMSE (rRMSE) are calculated in each regression technique. RMSE is one of the measures indicating how accurately the regression model predicts the dependent variable (in this study, economic losses) by calculating the residuals, which are the same as the difference between observed data and predicted value. Both training RMSE and validation RMSE are calculated to compare the accuracy

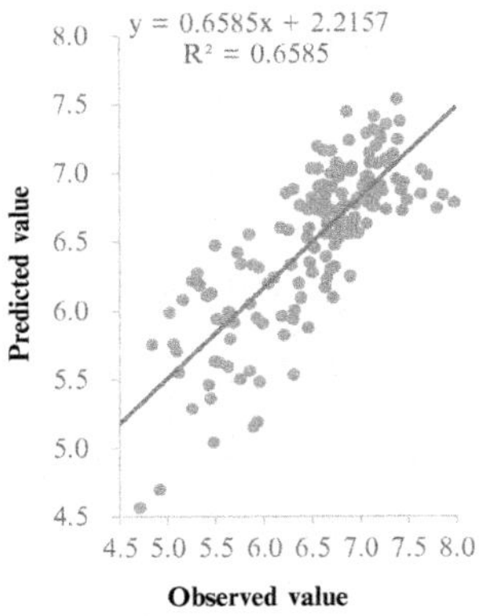

a. Linear regression using training data

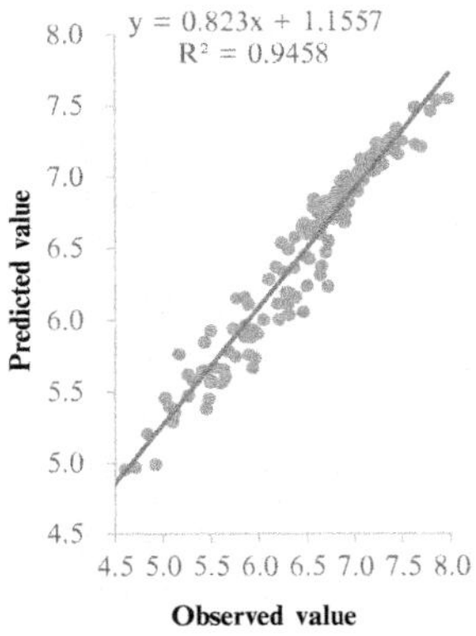

b. Random Forest modeling using training data

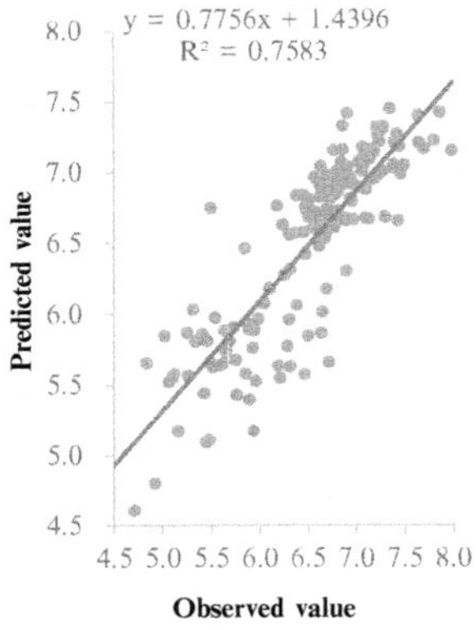

c. Cubist modeling using training data

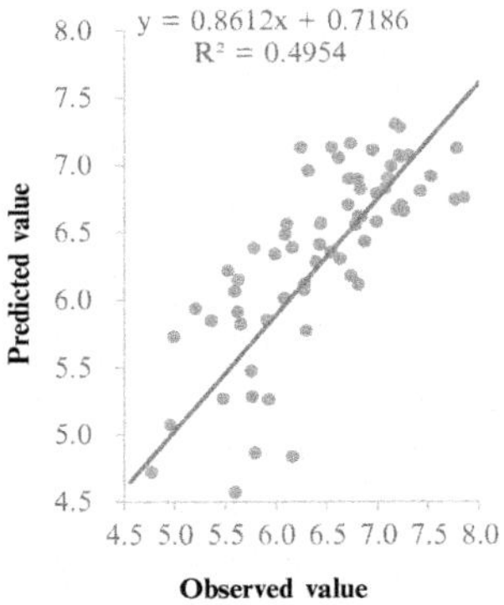

d. Linear regression using validation data

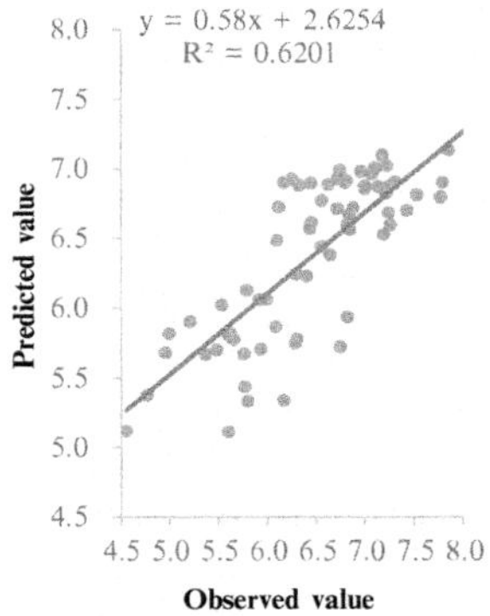

e. Random Forest modeling using validation data

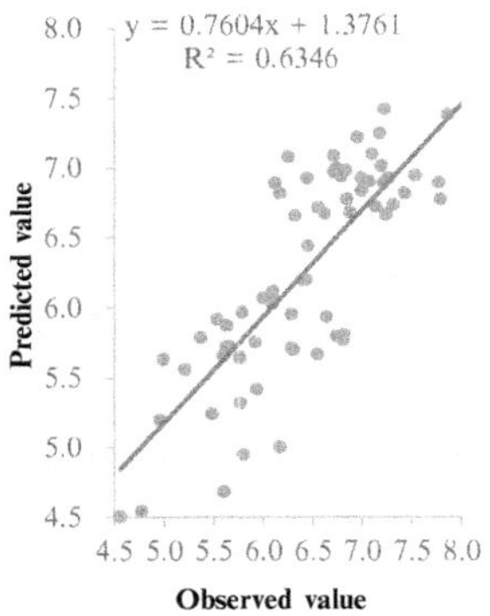

f. Cubist modeling using validation data

Fig. 4 Predicted value vs. observed values for each regression method

among the three regression techniques. The use of RMSE, however, has a limitation in that it cannot compare the error of different variables with different units (Richter et al. 2012). To address this limitation, this study calculated relative RMSE (rRMSE), which represents the ratio of RMSE to the mean of observed data for the training or validation data set.

As shown in Table 5, the results showed that machine learning techniques, both Random Forest and Cubist, produced lower RMSE and rRMSE (%) than the traditional OLS regression method in both training and validation data. The RMSE and rRMSE (%) of Random Forest were the lowest values (0.145 and 2.23 % for training data, 0.483 and 7.48 % for validation data) in both training and validation data.

Table 5 Summary of RMSE and rRMSE (%) for each regression method

		Linear regression	Random Forest	Cubist
Training	RMSE	0.291	0.145	0.264
	rRMSE (%)	4.48 %	2.23 %	4.06 %
Validation	RMSE	0.699	0.483	0.510
	rRMSE (%)	10.84 %	7.48 %	7.89 %

5 Discussion

Prior studies have noted the importance of assessing the vulnerability to natural disasters at a community level. As mentioned in the literature review, vulnerability assessment has been known as the methodology contributing to developing emergency plans (Uitto 1998), improving emergency management (Cutter and Finch 2008; Gao et al. 2014), developing and implementing mitigation strategies for damage reduction (Emrich and Cutter 2011), and identifying the causes of different levels of damage (Yoon 2012). In reviewing the literature, few studies on assessing vulnerability using various regression techniques have been found.

This study was designed to construct an index of disaster vulnerability of local communities in Korea to assess community vulnerability to natural disasters. First, this study deductively selected 12 vulnerability indicators including social, economic, and natural environment and built environment aspects based on previous vulnerability literature to examine the vulnerability of local communities in Korea. Among the selected indicators, important vulnerability indicators were examined and compared by the use of traditional OLS regression as well as two machine learning techniques, Random Forest and Cubist. Lastly, this study examined and compared the model performances and accuracy of each regression model by assessing the explanatory power and root-mean-square-error (RMSE).

The results of this study indicate that the three different regression techniques predicted similar important vulnerability indicators related to natural disasters. This finding is in close agreement with those in the earlier studies using machine learning approaches (Grömping 2009; Kim et al. 2014). Another important finding is that both training and validation regression models of machine learning techniques showed more powerful model performance in terms of R^2 and higher model accuracy in terms of RMSE and rRMSE (%) than that of OLS regression, which is consistent with the previous studies (Yoo et al. 2012; Kim et al. 2014).

This study has important implications for identifying the vulnerability indicators in Korea by applying not only the OLS regression method but also machine learning techniques, which have rarely been used in vulnerability studies. This study is expected to suggest the machine learning techniques that show powerful model performance as a tool for assessing vulnerability to natural disasters. Further research should be done to investigate the vulnerability indicators using machine learning approaches.

6 Conclusion

This study has assessed community vulnerability to natural disasters in Korea. The present study was undertaken to identify and compare the vulnerability indicators examined from three different statistical methods in 230 local communities in Korea.

One of the most significant findings to emerge from this study is that five characteristics out of 12, including industry density, precipitation, greenhouses, less educated people, and urban population, were found to be vulnerable characteristics in communities from three different regression models. A community with dense manufacturing establishments and a high portion of the population living in the urban area are less vulnerable to natural disasters than others. Contrary to the results of density of industry and urban population, a community with a high area of greenhouses shows higher vulnerability to natural disasters than others with less greenhouses. These results indicate that urban regions are less vulnerable to natural disasters than rural areas because manufacturing establishments are mainly located in the safe areas from natural disaster and greenhouses are concentrated in the risky area. A less educated people is a significant demographic factor increasing community vulnerability to natural disasters. From the natural environment aspects, the amount of annual precipitation is a statistically significant indicator increasing the disaster damage in a community.

The second major finding is that the regression models that use machine learning techniques showed better model performance and model accuracy than OLS regression. Although OLS regression is a simple technique to examine vulnerability, the violation of data assumptions weakens the model performance. Machine learning techniques, however, have no need for the data assumptions necessary to OLS regression, which results in better model performance than OLS regression. The findings of this study suggest the use of machine learning techniques in assessing community vulnerability to natural disasters as well as traditional OLS regression. The suggested methods used for this study may be applied to other countries' vulnerability assessment.

References

Armaş, I., and A. Gavriş. 2013. Social vulnerability assessment using spatial multi-criteria analysis (SEVI model) and the Social Vulnerability Index (SoVI model) – A case study for Bucharest, Romania. *Natural Hazards and Earth System Sciences* 13(6): 1481–1499.

Biass, S., C. Frischknecht, and C. Bonadonna. 2013. A fast GIS-based risk assessment for tephra fallout: The example of Cotopaxi volcano, Ecuador. *Natural Hazards* 65(1): 497–521.

Breiman, L. 2001. Random forests. *Machine Learning* 45(1): 5–32.

Burton, C., and S.L. Cutter. 2008. Levee failures and social vulnerability in the Sacramento-San Joaquin Delta area, California. *Natural Hazards Review* 9(3): 136–149.

Chang, S.E., J.Z. Yip, S.L.V.Z. de Jong, R. Chaster, and A. Lowcock. 2015. Using vulnerability indicators to develop resilience networks: A similarity approach. *Natural Hazards* 78(3): 1827–1841.

Cutter, S.L., and C. Finch. 2008. Temporal and spatial changes in social vulnerability to natural hazards. *Proceedings of the National Academy of Sciences of the United States of America* 105 (7): 2301–2306.

Cutter, S.L., B.J. Boruff, and W.L. Shirley. 2003. Social vulnerability to environmental hazards. *Social Science Quarterly* 84(2): 242–261.

Díaz-Uriarte, R., and S.A. De Andres. 2006. Gene selection and classification of microarray data using random forest. *BMC Bioinformatics* 7(1): 1.

Emrich, C.T., and S.L. Cutter. 2011. Social vulnerability to climate-sensitive hazards in the southern United States. *Weather, Climate and Society* 3(3): 193–208.

Esnard, A.M., A. Sapat, and D. Mitsova. 2011. An index of relative displacement risk to hurricanes. *Natural Hazards* 59(2): 833–859.

Felsenstein, D., and M. Lichter. 2014. Social and economic vulnerability of coastal communities to sea-level rise and extreme flooding. *Natural Hazards* 71(1): 463–491.

Fengqing, J., Z. Cheng, M. Guijin, H. Ruji, and M. Qingxia. 2005. Magnification of flood disasters and its relation to regional precipitation and local human activities since the 1980s in Xinjiang, Northwestern China. *Natural Hazards* 36(3): 307–330.

Flax, L.K., R.W. Jackson, and D.N. Stein. 2002. Community vulnerability assessment tool methodology. *Natural Hazards Review* 3(4): 163–176.

Gao, X., H. Yuan, W. Qi, and S. Liu. 2014. Assessing the social and economic vulnerability of urban areas to disasters: A case study in Beijing, China. *International Review for Spatial Planning and Sustainable Development* 2(1): 42–62.

Grömping, U. 2009. Variable importance assessment in regression: Linear regression versus random forest. *American Statistician* 63(4): 308–319.

Hair, J.F., R.E. Anderson, R.L. Tatham, and W.C. Black. 1998. *Multivariate data analysis with readings*. Upper Saddle River: Prentice Hall.

Hamilton, L.C. 1992. *Regression with graphics: A second course in applied statistics*. Belmont: Duxbury Press.

Im, J., J.R. Jensen, M. Coleman, and E. Nelson. 2009. Hyperspectral remote sensing analysis of short rotation woody crops grown with controlled nutrient and irrigation treatments. *Geocarto International* 24(4): 293–312.

Im, J., Z. Lu, J. Rhee, and L.J. Quackenbush. 2012. Impervious surface quantification using a synthesis of artificial immune networks and decision/regression trees from multi-sensor data. *Remote Sensing of Environment* 117: 102–113.

Kim, Y.O., S.B. Seo, and O.J. Jang. 2012. Flood risk assessment using regional regression analysis. *Natural Hazards* 63(2): 1203–1217.

Kim, Y.H., J. Im, H.K. Ha, J.K. Choi, and S. Ha. 2014. Machine learning approaches to coastal water quality monitoring using GOCI satellite data. *GIScience & Remote Sensing* 51(2): 158–174.

Korean Statistical Information Service. 2011. http://kosis.kr/. Accessed 22 Aug 2015.

Kuhlicke, C., A. Scolobig, S. Tapsell, A. Steinführer, and B. De Marchi. 2011. Contextualizing social vulnerability: Findings from case studies across Europe. *Natural Hazards* 58(2): 789–810.

Li, M., J. Im, L.J. Quackenbush, and T. Liu. 2014. Forest biomass and carbon stock quantification using airborne lidar data: A case study over Huntington wildlife forest in the Adirondack Park. *IEEE Journal of Selected Topics in Applied Earth Observations and Remote Sensing* 7(7): 3143–3156.

Liaw, A., and M. Wiener. 2002. Classification and regression by randomForest. *R news* 2(3): 18–22.

Malhotra, N.K., M. Peterson, and S.B. Kleiser. 1999. Marketing research: A state-of-the-art review and directions of the twenty-first century. *Journal of the Academy of Marketing Science* 27(2): 160–183.
Martins, V.N., D.S. e Silva, and P. Cabral. 2012. Social vulnerability assessment to seismic risk using multicriteria analysis: The case study of Vila Franca do Campo (São Miguel Island, Azores, Portugal). *Natural Hazards* 62(2): 385–404.
McEntire, D. 2012. Understanding and reducing vulnerability: From the approach of liabilities and capabilities. *Disaster Prevention and Management: An International Journal* 21(2): 206–225.
Minasny, B., and A.B. McBratney. 2008. Regression rules as a tool for predicting soil properties from infrared reflectance spectroscopy. *Chemometrics and Intelligent Laboratory Systems* 94 (1): 72–79.
Myers, C.A., T. Slack, and J. Singelmann. 2008. Social vulnerability and migration in the wake of disaster: The case of Hurricanes Katrina and Rita. *Population and Environment* 29(6): 271–291.
Perrow, C. 2011. *The next catastrophe: Reducing our vulnerabilities to natural, industrial, and terrorist disasters*. Princeton: Princeton University Press.
Richter, K., C. Atzberger, T.B. Hank, and W. Mauser. 2012. Derivation of biophysical variables from earth observation data: Validation and statistical measures. *Journal of Applied Remote Sensing* 6(1): 063557–1.
Runyan, R.C. 2006. Small business in the face of crisis: Identifying barriers to recovery from a natural disaster. *Journal of Contingencies and Crisis Management* 14(1): 12–26.
Schipper, L., and M. Pelling. 2006. Disaster risk, climate change and international development: Scope for, and challenges to, integration. *Disasters* 30(1): 19–38.
Tu, J., and Z.G. Xia. 2008. Examining spatially varying relationships between land use and water quality using geographically weighted regression I: Model design and evaluation. *The Science of the Total Environment* 407(1): 358–378.
Uitto, J.I. 1998. The geography of disaster vulnerability in megacities: A theoretical framework. *Applied Geography* 18(1): 7–16.
Yoo, S., J. Im, and J.E. Wagner. 2012. Variable selection for hedonic model using machine learning approaches: A case study in Onondaga County, NY. *Landscape and Urban Planning* 107(3): 293–306.
Yoon, D.K. 2012. Assessment of social vulnerability to natural disasters: A comparative study. *Natural Hazards* 63(2): 823–843.
Yu, D., and C. Wu. 2006. Incorporating remote sensing information in modeling house values. *Photogrammetric Engineering & Remote Sensing* 72(2): 129–138.
Zahran, S., S.D. Brody, W.G. Peacock, A. Vedlitz, and H. Grover. 2008. Social vulnerability and the natural and built environment: A model of flood casualties in Texas. *Disasters* 32(4): 537.
Zhou, Y., N. Li, W. Wu, and J. Wu. 2014. Assessment of provincial social vulnerability to natural disasters in China. *Natural Hazards* 71(3): 2165–2186.

Indirect Impact of Nuclear Power Plant Accidents Using an Integrated Spatial Computable General Equilibrium Model with a Microsimulation Module on the Korean Transportation Network

Euijune Kim and Younghyun John Kwon

Abstract The purpose of this paper is to assess the spatial and economic impact of the malfunction of nuclear power plants in Korea using a spatial computable general equilibrium (SCGE) model with a microsimulation module of the railroad and highway networks. This integrated approach takes into account the flows of commodities and input of factors among regions and industries using the spatial interaction and accessibility. The economic agents of the model consist of the producers, households, and governments from 16 Korean city-province regions, each having seven industrial sectors. While the microsimulation module of the highway and railroad networks measures the change in the regional accessibility of highways and railroads after the disaster, the SCGE model estimates the spatial effect of the reduction in growth potential and regional accessibility levels caused by the disaster on the economies. These counterfactual experiments show that accidents at nuclear power plants could lead to a reduction of GDP by 3.87 %, and the negative effects on the gross regional product (GRP) tend to become more severe in the Busan MA than any other regions by 3.03 %. The GRP levels in Seoul could decrease by 2.96 %, which is the smallest recorded amount from the six areas, due to its fewer economic interactions with the rest of Korea and a kind of reflexive benefit.

Keywords SCGE model • Unexpected events • Nuclear disaster • Economic impact analysis

E. Kim (✉)
Department of Agricultural Economics and Rural Development and Research Institute of Agricultural and Life Science, Seoul National University, Seoul, Korea
e-mail: euijune@snu.ac.kr

Y.J. Kwon
Department of Agricultural Economics and Rural Development, Seoul National University, Seoul, South Korea
e-mail: kwonyh98@snu.ac.kr

E. Kim, B.H.S. Kim (eds.), *Quantitative Regional Economic and Environmental Analysis for Sustainability in Korea*, New Frontiers in Regional Science: Asian Perspectives 25, DOI 10.1007/978-981-10-0300-4_8

1 Background

Unexpected damage from events such as typhoons, earthquakes, floods, and man-made disasters is expected to generate negative effects on economies, and their intensities and scopes tend to depend on the degrees of regional and industrial linkages and spatial networks. In particular, if the impact analyses are executed during the pre-disaster stage, the results can contribute to the implementation of policies that mitigate the damages for cost-effectiveness. For example, what is an optimal amount of a government budget in the precautionary stage? What are the direct and indirect economic benefits of disaster prevention strategies?

The purpose of this paper is to assess the spatial and economic impact of the malfunction of nuclear power plants in Korea using a spatial computable general equilibrium (SCGE) model with a microsimulation module of the railroad and highway networks by Kim and Yi (2015). This integrated approach takes into account the flows of commodities and input of factors among regions and industries using the spatial interaction and accessibility. The economic agents of the model consist of the producers, households, and governments from 16 Korean city-province regions, each having seven industrial sectors (agriculture and mining, information technology manufacturing, biotechnology manufacturing, nanotechnology manufacturing, mechanical technology manufacturing, construction, and services). While the microsimulation module of the highway and railroad networks measures the change in the regional accessibility of highways and railroads after the disaster, the SCGE model estimates the spatial effect of the reduction in growth potential and regional accessibility levels caused by the disaster on the economies. This counterfactual experiment assumes that the nuclear power plants in the eastern area of Korea are damaged by natural disasters, consequently causing an accident that includes the loss of coolants and release of radioactive materials after the nuclear meltdowns. This process is similar to the 2011 Fukushima Daiichi nuclear disaster in Japan and the 1986 Chernobyl disaster in Russia. This paper is divided into three sections. The next section is focused on literature reviews on the application tools that analyze the economic impact of the unexpected events on the regions. The third section measures the economic impact of the disasters on the regional economies, while discussing a structure of the integrated SCGE model with the transportation module. The final section provides a summary of the research and further research issues.

2 Literature Review

There have been numerous attempts to analyze and measure the regional economic impact of natural disasters. The most popular approach to the economic analysis of disasters is based on the input-output (IO) model. Ryu and Cho (2010) estimated the overall economic losses due to typhoons and heavy rains in Korea and developed an "event matrix," which was designed to calculate the new input-output structure after the natural disaster. Rose et al. (1997) estimated the regional economic losses from earthquake-damaged electric utility lifelines in the New Madrid Seismic Zone in Tennessee in the USA from IO and linear programming models. According to them, the potential production losses over the recovery period could amount to as much as 7 % of GRP of the state. They showed how losses could be reduced by reallocating electricity resources and optimizing their recovery sequences through linear programming. Okuyama (2004) examined the impact of the 1995 Great Hanshin earthquake in Japan and its recovery process on the dynamic processes of the regional economies by using Miyazawa's extended input-output framework and sequential interindustry model (SIM). The SIM framework was originally developed to analyze interindustry production in a dynamic economic environment, such as in large construction projects where the effects on production and employment are transitory. Gordon et al. (1998) also applied the Southern California Planning Model and the IO-based model to the calculation of the costs of business interruption during the 1994 Northridge earthquake. Their analysis found that the interruption accounted for 25–30 % of the full costs of the earthquake.

There have been additional efforts to integrate a transportation network with the IO model, such as Cho et al. (2001) and Sohn et al. (2003). Sohn et al. (2003) analyzed the economic impact of the earthquake on the transportation network for the Midwest states, taking into account a loss of final demand and an increase of transport cost. The modeling system included a transportation network loss function, a final demand loss function, and an integrated commodity flow model. They showed that the economic significance was determined by not only the level of disruption but also the volume of flow on the link, the relative location (topology) on the highway network, and the economic intensity near the network link. Cho et al. (2001) developed an integrated and operational system to explore how the economic losses of the transportation services and industrial capacities from the earthquake affected the Los Angeles economy. The composited approach was composed of a performance model of integrated bridges and other structures, a model of the transportation network, a model of the spatial allocation, and an IO model. They found that the spatial distribution of these losses was sensitive to changes in network costs by the disruption of transportation services.

The CGE is another analytic framework that measures the economic impact of natural disasters. Bosello et al. (2007) measured the amounts of land and capital losses due to a sea level rise in coastal regions by applying the CGE model to experiments on coastal area protection. They found that the GDP and energy consumption in regions that substantially build dikes could increase, in spite of a reduction in the utility level under coastal protection. Tatano and Tsuchiya (2008) developed an analytical framework to estimate the economic losses by the disruption of the transportation network after the Niigata-Chuetsu earthquake, finding that the economic costs relied on interregional trade patterns. Rose and Liao (2005) analyzed the economic impact of a disruption of water services in the Portland metro economy. They showed how indirect economic losses varied depending on the overall level and sectoral mix of water shortages, the extent of pre-event mitigation, and the post-event inherent and adaptive resilience.

For the econometric model, Padli et al. (2010) carried out a cross-sectional analysis to examine a relationship between the impact of natural disasters on economies and their socioeconomic backgrounds. The major determinants for the values of the impact included GDP per capita, the ratio of government consumption to GDP, education, the land area, and the size of the population. Richard et al. (1984) developed a regional econometric model to assess the potential effects of the disaster on the economy. It was interesting to use supply-side factors such as capital investment, migration, and transportation as the independent variables, since the demand-side ones have been prevalently used in traditional models. They argued that the economic impact from natural disasters depended on the degree of spatial disaggregation. This paper showed that reconstruction was a key factor on the long-run growth and recovery path of regions in the sense that the income gains from the post-disaster period could offset the income loss by the disaster. Hong et al. (1996) estimated the monetary costs incurred by accidents in the nuclear power plants, while the overall costs were disaggregated into replacement power costs, capital investment costs, plant repair costs, early decommissioning costs, health-care costs, evacuation costs, relocation costs, disposal of agricultural product costs, and decontamination costs.

There are a few difficulties in assessing the economic impact of natural disasters, as West and Lenze (1994) pointed out. First of all, the size of any event is uncertain and unknown. It is not easy to classify the economic loss of the disasters into industrial sectors for the economic analysis, because doing so could cause a multiple-counting problem to measure the benefits and costs of the disasters. In addition, it is difficult to capture variations in the economic behavior of households and firms after natural disasters, and the final outcomes from the analysis might be sensitive to initial conditions and assumptions underlined in the model. Regarding the analytical tools, the IO model and its extended applications, such as the event matrix, have been popular methods for impact analysis, but most of them deal with only the short-run economic consequences. They need to incorporate a traditional demand-side approach with the supply-side one. In addition, though the CGE model can quantify the economic effects in terms of prices as well as quantity sides, it also requires the development of an operational mechanism to identify dynamic changes in the socioeconomic behavior of economic agents from the disaster to the recovery point in a systematic way (Table 1).

Table 1 Method for an impact analysis of disasters

Author	Type of disasters	Model	Impacts/key issues
Richard et al. (1984)	Earthquake	Econometric model	Long-run and short-run impacts
Hong et al. (1996)	Nuclear plant disruption	Cost analysis	Cost analysis
Rose et al. (1997)	Earthquake	Input-output model and linear programming	Reduction in GRP (7 %)
Gordon et al. (1998)	Earthquake	Input-output model	Cost of business interruption
Cho et al. (2001)	Earthquake	Input-output model	Integration of network model, spatial allocation model, and IO model
Narayan (2003)	Cyclone	CGE model	Macroeconomic impacts of cyclone "Ami" for Fiji Islands
Sohn et al. (2003)	Earthquake	Input-output model	Network effects on transportation
Okuyama (2004)	Earthquake	Sequential interindustry model	Impacts on interregional and interindustrial sectors
Rose and Guha (2004)	Earthquake	CGE model	Impacts of the electric utility lifeline disruptions
Rose and Liao (2005)	Disruption in water service	CGE model	Impacts of water service disruptions
Bosello et al. (2007)	Sea level rise	CGE model	Impacts on GDP and energy consumptions
Tatano and Tsuchiya (2008)	Earthquake	SCGE model	Direct and indirect spillover effect on regional economies
Ryu and Cho (2010)	Typhoon and heavy rain	Input-output model	Reduction in GDP (1.18 %)
Padli et al. (2010)	Natural disaster	Econometric model	Linkage between economic condition and impacts

3 Analysis

3.1 Integrated SCGE Model

Since the major sources of damage from disasters are disruptions in transportation networks and reductions in production activities, we develop a systematic model, namely, an integrated SCGE model, based on two sub-models: a transport model and an SCGE model. The former model is designed to explore how the disaster affects the levels of spatial accessibility of highways and railroads. The accessibility for each mode is derived from the minimum travel times among 237 city and county zones. "Accessibility" refers to the level of services provided by transportation networks and, by definition, is determined by the population or employment by origin and destination and minimum time between two nodes. The travel time is derived from the shortest route algorithm in the ARC-GIS with an average speed on each link by mode.

The SCGE model estimates the impacts of the disaster across the entire economy by dividing it into 16 economic provinces and using changes in the levels of spatial accessibility and factor inputs, such as capital stocks and labor demands. The SCGE model is spatially disaggregated into 16 domestic provinces, as well as one that represents the rest of the world (ROW). In each domestic province, the production activity is classified into seven industrial sectors: agriculture and mining, information technology manufacturing, biotechnology manufacturing, nanotechnology manufacturing, mechanical technology manufacturing, construction, and services. The mathematical structure of this model is based on the work done by Kim and Yi (2015) and Kim et al. (2004). The SCGE model accounts for the economic behavior of producers and consumers on the real side economy, following the approach of neoclassical elasticity including market-clearing prices, the maximization of a firm's profits, and a household's utility. The economic agent selects an optimal set of factor inputs or commodity demands under the maximization principles of constrained profit or private utility. The regional products are then disaggregated into intraregional demands, regional imports, and foreign imports in terms of the product origin or intraregional supplies, regional exports, and foreign exports in terms of the product destination. Each price, including the commodity price and factor input price, is adjusted to balance between supply and demand in the market.

Our production structure is multilevel. The gross output by region and sector is determined via a two-level production function of value-added and composite intermediate inputs. The intermediate inputs are derived from input coefficients, whereas the value-added element is determined by a translog production function of labor and capital inputs, and the spatial accessibility variable as a proxy for the level of transportation services. The labor demand is derived from the value-added maximization of the first-order conditions of producers, and the labor participation rate is derived by balancing out total labor demand against total labor supply under the neoclassical closure rule for the labor market. The in-migration is assumed to be in response to interregional differences between origin and destination regions in terms of wage per capita and unemployment rate, as well as the physical distance between the regions. The cost minimization from the Armington approach accounts for an optimal ratio of the foreign import to the domestic sale, while the latter is disaggregated into demands for 16 regional goods under the Cobb-Douglas function. The profit maximization with the two-level constant elasticity of transformation function determines an optimal allocation of the gross output into the foreign export and domestic supply. According to the concept of commodity equilibrium, the latter includes both intraregional supplies and regional exports.

The total demand for goods and services consists of intermediate demands, total consumption expenditures for households, government consumption expenditures, and investments. The household income consists of wages, capital income, and

Table 2 Major equations from the SCGE model

Output	Output = Leontief (value added, intermediate demand)
Value added	Value added = total factor productivity*CD (capital stock, labor)
Supply	Output = CET (foreign exports, domestic supply)
Domestic supply	Domestic supply = CET (regional exports, intraregional supply)
Demand	Demand = Armington (foreign imports, domestic demand)
Domestic demand	Domestic demand = CD (regional imports, intraregional supply)
Labor demand	Labor demand = LD (wage, value added, net price)
Total factor productivity	Total factor productivity = TFP (accessibility, population)
Labor supply	Labor supply = LS (labor market participation rate, population)
Population	Population = natural growth of previous year's population + net population inflows
Regional incomes	Regional incomes = wage + capital returns + government subsidies
Migration	Migration = TODARO (incomes and employment opportunities of origin and destination, distance between origin and destination)
Consumption by commodity	Consumption by commodity = CC (price, population size by age cohort, incomes)
Private savings	Private savings = PS (saving rate, income)
Government revenues	Government revenues = indirect tax + direct tax + tariff
Government expenditures	Government expenditures = government current expenditure + government savings + government investment expenditure + government subsidies
Labor market equilibrium	Labor demand = labor supply
Capital market equilibrium	Private savings = total investments
Commodity market equilibrium	Supply of commodities = demand for commodities
Government	Government expenditures = government revenues
Capital stock	Capital stock = depreciated lagged capital stock + new investments

Source: Kim et al. (2013)

exogenous subsidies from the government. Two tiers of government structure, 16 regional governments and one national government, are specified in the model. With regard to the macroeconomic closure rule for the capital market, total investments are determined by aggregate savings including household savings, corporate savings of regional production sectors, private borrowings from abroad, and government savings. The price adjustment is required for the Walrasian equilibrium condition, and the numeraire of the model is set as the consumer price index. In addition, we calibrate a social accounting matrix (SAM) as a benchmark for the development of the SCGE model. The SAM consists of six accounts—factors, households, production activities, government, capital, and the rest of the world—and is treated as an initial equilibrium for the SCGE model (Table 2).

3.2 Simulation

In the counterfactual experiment in this paper, we suppose that nuclear and radiation accidents, such as a release of critical amounts of radioactivity and damage to a reactor core, have occurred in Busan and Gyeongju, which is located in the eastern area of the country. These two locations represent 50 % of total number of nuclear power plants in Korea. Additionally, one of the nuclear power plants in Busan has not been operational due to a handling mistake, while more than 75 % of the malfunctions at these power plants have taken place in facilities that have operated for more than 20 years.

Since the analysis in this paper is concerned with estimating the indirect effects of disasters on regional economies, we do not take into account operation costs of the power plants as well as environmental and damage costs in the computation. The indirect effects are measured in terms of gross domestic product (GDP) and gross regional product (GRP) under free interindustrial mobility of labor and capital factors across regions. We assume that any production and transportation activities are not allowed within a 30 km radius of the damaged area for 6 months, based on the fifth level of the International Nuclear Event Scale. This scenario mimics the cases of Fukushima in Japan in 2011 and Chernobyl in Russia in 1986. There are two external shocks in this experiment: (1) a rate of 50 % of temporary production capacity loss of the manufacturing sectors and (2) completely blocking the use of highways and railroad network systems in the damaged areas for 6 months. We suppose that the reductions on the production potential and mobility of economic resources within the damaged areas result in a direct interruption of industrial activities, which affects the prices and quantities of other regional commodities and factor inputs through the multiregional interindustrial linkages of productions and consumptions. In other words, the economic outcomes from the disaster in this paper could depend on the degrees of the industrial specialization and linkages and the levels of spatial accessibility of the highway and railroad networks. The levels of the GRP from the baseline without the shock, and the counterfactual case with the shock, may depend on the underlying assumptions and types of exogenous variables in the model (Fig. 1).

For the analysis, 16 provinces are classified into five metropolitan areas (MA) in Seoul, Daejeon, Gwangju, Daegu, and Busan and one remote region in Gangwon and the Jeju area, while the damaged areas, Busan and Gyeongju, are in the Busan and Daegu metropolitan areas, respectively. Two external shocks on these regions could lead to an expected downturn in the national and regional incomes. The GDP as a total amount of the GRP declines by 3.87 % when compared to the baseline, as is shown in Table 3, and the negative effects on the gross regional product (GRP) tend to become more severe in the Busan MA than any other regions; the growth rate of the GRP is –6.00 %, which is lower than that of the Daegu MA, another damaged area, by 0.87 %. The disaster decreases the GRP of two remote areas, Gangwon and Jeju, by 3.87 %, as well. However, a somewhat interesting point is that the GRP levels of Seoul decrease by 2.96 %, the lowest level of the six areas

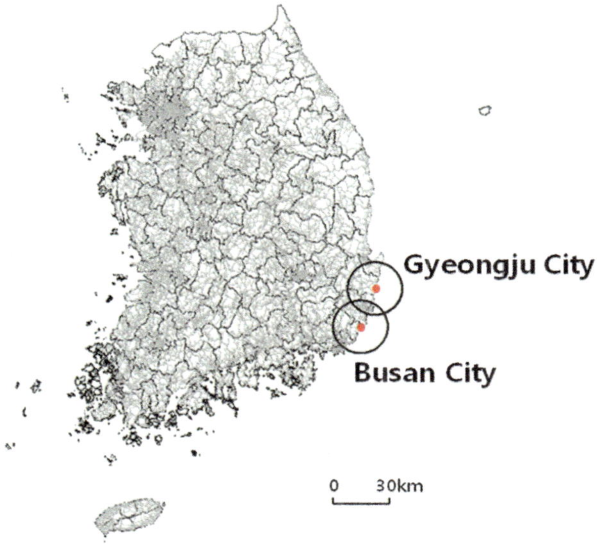

Fig. 1 Damaged areas by nuclear power plant accidents

tested. One of reasons for this result is that the Seoul MA has less economic connections with the damaged areas in terms of production flows. Another is that the Seoul MA might enjoy "reflexive benefits" by increasing their market shares in the domestic markets for some time, while other regions are experiencing regional income losses from the accidents.

Since the disaster shocks are composed of reductions in the production capacity and levels of spatial accessibility of highways and railroads, the impact on the GDP and GRP can be structurally decomposed into three components: (1) the first term, the income variation from changes in only production capacity; (2) the second term, the income variation from changes in only spatial accessibility; and (3) the final term, the income variation from the interaction between changes in production capacity and accessibility as a residual.

Net Effects
= Income levels from the baseline (without the shock) – income levels from the counterfactual case (with the shock)
= (1) Income variation from changes in only production capacity
+ (2) Income variation from changes in only spatial accessibility
+ (3) Income variation from the interaction between changes in production capacity and accessibility (1 and 2) as a residual

It is possible to trace out the income changes by each disaster source; the loss in capital stock in the damaged areas has a negative effect on the GDP of −2.88 %, which is 74.42 % of the total effect (−3.87 %) in column A of Table 3. The impacts on the national economy by a reduction in the accessibility and interaction of two damage sources are −0.71 % (18.35 % of the total) and −0.28 % (7.23 % of the

Table 3 Simulation results on gross regional product (unit, %)

Macro-region	Province	Type	Mid-long term	Long term	Long term
Assumption on labor market			Mobile between provinces	Mobile between provinces	Mobile between provinces
Assumption on capital market			Mobile within provinces	Mobile between provinces	Mobile between provinces
Capital area	Seoul	Capital	−2.45	−1.64	−9.32
	Inchon	Largest city	−4.20	−4.94	−11.16
	Kyunggi	–	−2.79	−4.30	−9.25
Central area	Daejeon	Largest city	−3.31	−5.10	−11.76
	Chung-buk		−3.95	−5.28	−11.77
	Chung-nam		−4.25	−5.41	−11.71
Western area	Kwangju	Largest city	−3.13	−4.83	−11.10
	Jeon-buk	–	−3.20	−4.81	−11.33
	Jeon-nam	–	−3.49	−4.48	−10.38
Eastern area	Daegu	Largest city	−5.12	−4.43	−11.37
	Kyung-buk		−11.94	−8.52	−17.39
	Busan	Largest city	−6.39	−5.89	−13.78
	Ulsan	Largest city	−15.57	−12.40	−23.61
	Kyung-nam		−6.49	−6.77	−14.83
Mountain area	Kangwon	–	−2.71	−4.83	−11.52
Island	Jeju	–	−3.54	−5.18	−12.44
Average			−4.72	−5.37	−11.81
Coefficient of variation			−0.76	−0.43	−0.30

total), respectively. This shows that if we measure the effects without linkages of the transportation network with the spatial mobility and allocation of the resources, the result could be underestimated. The Busan MA and Daegu MA have a large share of the economic impact of the disaster from the capacity loss, which measured as much as 83.00 % and 81.99 % of the total, respectively, while having relatively small shares from the interaction component as shown in column C of Table 3.

4 Conclusion and Further Research Agenda

This paper estimates the effects of unexpected disasters on regional economies by developing an integrated SCGE model. The results show that accidents at nuclear power plants could lead to a reduction of GDP by 3.87 %, and the negative effects on the gross regional product (GRP) tend to become more severe in the Busan MA than any other regions by 3.03 %. The GRP levels in Seoul could decrease by 2.96 %, which is the smallest recorded amount from the six areas, due to its fewer economic interactions with the rest of Korea and a kind of reflexive benefit.

Concerning agendas for further research, future efforts need to develop a dynamic model to calibrate the effects of disasters across the entire economy in both the short and long run. The model type should be disaggregated into a backward looking, or recursive type, and a forward looking, or perfect foresight type, depending on the assumption of the optimization behavior of the economic agents. The simulation on the evolution of the effects over time can be used to implement public infrastructure plans to minimize the total costs of a disaster, which includes the costs from the damage and investment expenditures for the recovery and reconstruction during the post-hazard stage (Okuyama and Santos 2014). These simulation results can also contribute to finding an optimal allocation of the government budget and assess long-term disaster plans and strategies in terms of a cost-benefit analysis. For instance, Sohn et al. (2003) analyzed the economic impact of an earthquake on the transportation network in the Midwest states. The modeling system was composed of the transportation network loss function, the final demand loss function, and the integrated commodity flow model. These could provide information to support a decision-making process on the optimal retrofit priority of bridges and links for a transportation network, by identifying the most significant link in a network, in an economic sense, and the link with the greatest physical disruption. Another possible effort in the future would have an emphasis on the estimation of the indirect effects on the households and producers in the non-damaged areas, as well as the direct effects on them in the damaged area. The former may involve a fundamental change in the economic structure to reflect the different economic behavior of the agents in the commodity market. For example, the national government announced the economic loss caused by the Middle East respiratory syndrome (MERS) outbreak was 9.8 billion US dollars as of June of 2015. Most of these losses were relevant to the indirect negative effects on the tourism and service industries, rather than the direct damage to human beings in the form of death and injuries.

References

Bosello, F., R. Roson, and R.S.J. Tol. 2007. Economy-wide estimates of the implications of climate change: Sea level rise. *Environmental & Resource Economics* 37: 549–571.

Cho, S., P. Gordon, J.E. MooreII, H.W. Richardson, M. Shinozuka, and S. Chang. 2001. Integrating transportation network and regional economic models to estimate the costs of a large urban earthquake. *Journal of Regional Science* 41(1): 39–65.

Gordon, P., H.W. Richardson, and B. Davis. 1998. Transport-related impacts of the Northridge earthquake. *Journal of Transportation and Statistics* 1: 21–36.

Hong, D., K. Lee, and J. Jeong. 1996. A study on the estimation of economic consequence of severe accident. *Proceedings of the Korean Nuclear Society autumn meeting*, 409–414.

Kim, Euijune, and Yoojin Yi. 2015. Impact of railroad investments on regional economies: An approach of spatial CGE model with a microsimulation module of railroad and highway networks. The 5th international seminar on integration of spatial computable general equilibrium and transport modelling. Kobe, Japan, October, 15–17.

Kim, Euijune, G.J.D. Hewings, and C. Hong. 2004. An application of integrated transport network multiregional CGE model I: A framework for economic analysis of highway project. *Economic System Research* 16(3): 235–258.

Kim, Euijune, G.J.D. Hewings and C. Lee. 2013. *University investment expenditures and economic loss from population aging in Korea*. The 3rd Asian Seminar in Regional Science, Hualien, Taiwan, August 7–8, 2013

Narayan, P.K. 2003. Macroeconomic impact of natural disasters on a small island economy: Evidence from a CGE model. *Applied Economics Letters* 10: 721–723.

Okuyama, Y. 2004. Modeling spatial economic impacts of an earthquake: Input-output approaches. *Disaster Prevention and Management* 13(4): 297–306.

Okuyama, Y., and J.R. Santos. 2014. Disaster impact and input-output analysis. *Economic Systems Research* 26(1): 1–12.

Padli, J., M.S. Habibullah, and A.H. Baharom. 2010. Economic impact of natural disasters' fatalities. *International Journal of Social Economics* 37(6): 429–441.

Richard, W.E., J.W. Milliman, and R.B. Roberts. 1984. Measuring the regional economic effects of earthquakes and earthquake predictions. *Journal of Regional Science* 24(4): 559–579.

Rose, A., and G.S. Guha. 2004. Computable general equilibrium modeling of electric utility lifeline losses from earthquakes. In *Modeling the spatial economic impacts of natural hazards*, ed. Yasuhide Okuyama and Stephanie Chang, 119–142. Heidelberg: Springer.

Rose, A., and S. Liao. 2005. Modeling regional economic resilience to disasters: A computable general equilibrium analysis of water service disruptions. *Journal of Regional Science* 45(1): 75–112.

Rose, A., J. Benavides, S.E. Chang, P. Szczesniak, and D. Lim. 1997. The regional economic impact of an earthquake: Direct and indirect effects of electricity lifeline disruptions. *Journal of Regional Science* 37(3): 437–458.

Ryu, M., and S. Cho. 2010. A study on national economic impact by natural disaster: A case of typhoon. *Productivity Review* 24(3): 273–290.

Sohn, J., T. Kim, G.J.D. Hewings, J. Lee, and S. Jang. 2003. Retrofit priority of transport network links under an earthquake. *Journal of Urban Planning and Development* 129(4): 195–210.

Tatano, H., and S. Tsuchiya. 2008. A framework for economic loss estimation due to seismic transportation network disruption: A spatial computable general equilibrium approach. *Natural Hazards* 44: 253–265.

West, C.T., and D.G. Lenze. 1994. Modeling the regional impact of natural disaster and recovery: A general framework and an application to Hurricane Andrew. *International Regional Science Review* 17(2): 121–150.

A New Framework to Quantifying the Economic Impacts of Cyberattacks on Aviation Systems: A Korean Game-Theoretic Interregional Economic Model

JiYoung Park, Minsu Son, Ha Hwang, Dongin Cho, and Changkeun Park

Abstract This study suggests a framework quantifying a cyberattack on the Korean airport security system. Recent cyberattacks on nuclear power plants in South Korea and a serious cyberattack on Sony Pictures in the USA indicate possible invasion to an airport electronic system because the invasion would not have any border or entry point. Korea governments must consider this complex process that may cause turmoil once occurred. This complicated situation highlights the need for improved intergovernmental collaboration within the Korean territory. First, to avoid cyberterrorist threats, it is essential to coordinate intergovernmental network closely. Designing this network should consider delivering both a competitive game between attackers and defenders and a cooperative game between governments. Second, because an airport shutdown that is only located in a certain area would cause ripple impacts throughout other domestic (and international) economies, the Korean interregional input–output model should be combined to capture this type of ripple impacts. A new framework suggested as the Korean game-theoretic interregional economic model contributes to understanding strategies in cyberterror security and identifying the probabilistic economic costs on a South Korean airport closure by place of event and by type of industry. Using the identified equilibrium strategies for the Korean airport protection, a general guideline to evaluate resource allocation can be passed onto the South Korean government agencies.

J. Park (✉) • H. Hwang • D. Cho
Department of Urban and Regional Planning, University at Buffalo, The State University of New York, Buffalo, NY, USA
e-mail: jp292@buffalo.edu; hahwang@buffalo.edu; didi9535@buffalo.edu

M. Son
Program in Regional Information, Department of Agricultural Economics and Rural Development, Seoul National University, Seoul, South Korea
e-mail: msson@snu.ac.kr

C. Park
Department of Agricultural and Applied Economics, University of Georgia, Atlanta, GA, USA
e-mail: beinji2@gmail.com

E. Kim, B.H.S. Kim (eds.), *Quantitative Regional Economic and Environmental Analysis for Sustainability in Korea*, New Frontiers in Regional Science: Asian Perspectives 25, DOI 10.1007/978-981-10-0300-4_9

Keywords Cyberattack • Airport security • Game theory • Interregional input–output model • The Korean game-theoretic interregional economic model

1 Cybersecurity and Terrorism in Korea

As a high-developed information technology (IT) society, South Korea has been benefited of various, advanced IT systems such as the Internet of Things (IoT) and ubiquitous and network infrastructure. Especially, the smart age generation prefers to use IoT and live in a big or megacity because their everyday lives should be connected by an online network to all around the Korean and other global territories (Park 2015). While they feel living a technologically advanced city can provide higher-quality and more attractive living conditions, paradoxically, these Internet-based network systems will become more vulnerable from cyberattacks as they get more closely connected. Considering that most public infrastructure and financial institutes are concentrated only to 16.6 % of all territories of South Korea where about 91.6 % of the South Korean population live (MOLIT 2014), cyberthreats become recognized as more serious fears to the entire Korean territory than ever.

However, little has been reported what economic, social, and/or culture impacts could be generated from these successful cyberattacks. It is easy to imagine that if major infrastructure such as Incheon International Airport (ICN) would have been attacked, the economic damages associated with the event would not be countable. Increasing virus threats with which hackers intended to disconnect various Internet service types from both numerous private and public institutes leads to developing a higher cybersecurity level. Once a cyberattack has succeeded, the economic damage may go beyond our expectation. Therefore, this study is to focus on providing a theoretical framework quantifying economic impacts of the potential damages. An issue is how to combine the successful probability of a cyberattack with an economic tool to be applied for measuring economic impacts both at the local and national levels.

Many studies have evaluated physical damages of terrorist attacks based on the potential economic consequences stemming from infrastructure damages and business interruptions in the affected regions of the USA. (Park 2008; Park et al. 2007; Richardson et al. 2007; Gordon et al. 2007; Richardson et al. 2014, 2015). Especially, Gordon et al. (2007) applied an input–output approach to measure the economic disruption of major US airports. This study provided us with useful information on the possible catastrophes of the US aviation system, analyzing the 9/11 events. Based on expenditures to mitigate and respond to emergent disruptions of the aviation system, various decision-makers could be advised to prepare future damages. However, it is very rare to find an economic impact study of a cyberevent on an aviation system.

A cyberinvasion of an airport electronic system by terrorists would require complex strategic behaviors. At the same time, any defending entity must consider the turmoil that may be caused by complex processes. The possibility of simultaneous cyberattacks further raises the difficulty for defending entities in securing their airports. This highlights the need for improved, integrated collaboration between local governments and between countries subject to this type of attack. Collaborative networking, connected domestically and internationally in terms of governmental cooperative integration, requires close intergovernmental coordination to overcome cyberterrorist threats. For example, if cyberterrorists could successfully invade an airport information system, they could cause problems in the operational software and the database that contains valuable information. If the airport is an international hub, a single event could affect not only the region that the airport is located in but also other domestic and international airports that are connected to that airport. According to Oxford Economics (2008), the air transport industry contributes substantially to the global economy—up to $1540 billion annually—and comprises 33.3 million jobs on average. Constructing a new model for strategic cyberterror security requires a combination of both competitive and cooperative game situations. This needs to develop specific strategies against cyberterrorism. In addition, an airport shutdown would have spillover effects throughout local and domestic economies. It is necessary to analyze these effects with a spatially disaggregate economic model.

How to combine the probability of cyberattacks with the consequent economic impacts? A new framework development needs to involve quantifying the possible economic impacts of breached strategic airport security. This leads to suggest a new probabilistic economic impact model that provides probabilistic estimates of differentiated economic impacts by region and industry. Furthermore, it is also important to consider both competitive and cooperative game situations with the economic impact model.

This study proposes a new framework of a game-theoretic economic model using the Korean interregional input–output model, naming the Korean game-theoretic interregional economic model (K-GIEM). Using K-GIEM, we can identify which airport would be most vulnerable in the event that an airport electronic system would be shut down. Calculating the probabilistic costs of airport closure, the K-GIEM determines the economic importance of cybersecurity by event location and industry type. Advancing in understanding of how cyberattacks affect the real economies of South Korea, the K-GIEM can identify equilibrium strategies for airport protection of South Korea against cyberterrorists. Also, it can provide a general guideline to evaluate resource allocation strategies for local government agencies of South Korea. Eventually, the framework suggested here will contribute to providing a basis on aviation security and policy for communication among policy-makers, the general public, and local economic entities.

The rest of this study is organized as follows. The next section discusses relations among cyberattacks, strategic behaviors, and the consequent damages. Section 9.3 introduces both competitive and cooperative strategic behaviors involved in a game-theoretic situation. Section 9.4 focuses on an interregional input–output (IRIO) model released by the Bank of Korea. Section 9.5 explains how an integrated modeling framework, the Korean game-theoretic interregional economic model (K-GIEM), can involve behavioral strategies of cyberattack and the security response with the Korean IRIO model, which is needed for measuring probabilistic economic costs. The final section concludes this study with a brief summary, discussing further research as well.

2 Cyberattacks, Strategic Behaviors, and Consequent Damages

Massive distributed denial of service (DDoS) attacks have been occurring in many fields of South Korea. As demonstrated in Table 1, the cyberattacks occurred in South Korea can be categorized into two types: data hacking and infrastructure disruption. Regarding the data hacking, a big data analysis that may provide new information through the collected data can be an effective, dangerous weapon to attackers. If information of a Korean financial institute outflows, the damage is not ceased within South Korea because the financially connected network is already globalized and complicated enough to restore the system to the previous status undamaged. While several hackings on financial institutes in South Korea have not been reported to make direct damages, it is also unknown of how new information collected from the hacked data has been protected and clearly safe without making any negative effect.

Second, physical facility threats may lead to a catastrophe due to the high-dense society that South Korea has. In August in 2014, 53 official, confidential documents of Korea Railroad Corporation (KORAIL) were leaked from an outside attack that was based on a network structure method. Especially, because a train network system has been developed to a narrow territory of South Korea and is a very popular transportation mode that connects all around the Korean territory, it is expected that various direct and indirect damages would occur. Also, a nuclear power plant explosion in Fukushima, Japan, occurred from a cooling system error by electronic blackout; the damages have been emerging throughout the world. In December 2015, even though any physical damage has not been reported, a technology hacking on nuclear power generation in South Korea was reported as a very dangerous threat because any tiny disruption of the system that may affect the cooling system could generate irreversible huge damages on the entire nation.

However, this cyberissue is not a local issue restricted to Korea anymore. For example, cyberhackers may be able to disable substantial public transportation control systems because modern public transportation network systems that largely rely on computerized systems can be invaded without spatial and temporal borders (Koscher et al. 2010; Ignelzi 2012). Indira Gandhi International Airport experienced a simple technical failure in the Common Use Passenger Processing System in 2011. This is reported as a result of a virus attack on the system, generating system shutdown for half a day and delaying about 50 flights for 15–20 min (Kakkar 2011). More recently, the USA also experienced a cyberattack of a malware insertion on airport information systems that might have caused serious

Table 1 Various successful cyberattacks occurred in South Korea

Date	Target	Form of attack and symptom
7.7.2009	The Blue House, the National Assembly in S. Korea, the US Department of the Treasury, and the US Department of Homeland Security	DDoS attack for computer network breakdown
3.4.2011	The Blue House, various national institutes including National Intelligence Service (NIS), financial institutes including Kookmin Bank, and several Internet companies including NAVER	DDoS attack for computer network breakdown
4.12.2011	Nonghyup	Malicious code infection for service interruption
6.9.2012	The JoongAng Ilbo	Home page falsification to delete data
3.20.2013	The press including KBS, MBC, and YTN and financial institutes including Shinhan Bank and Nonghyup	Computer network breakdown
3.25.2013	Nalssi.com	Spread a malicious code for personal computer infection
3.26.2013	Local autonomous entity computing center, the Ministry of Strategy and Finance home page, and YTN home page. It seems related to North Korea organization home page	Computer network breakdown
8.7.2014	KORAIL	Technology document hacking and network structure method hacking
12.9.2014	Korea Hydro & Nuclear Power Co.	Nuclear power generation technology hacking

Source: Yonhap News on 06.25.2013 and revised by authors (accessed on 09.22.2015)

malfunction in diverse systems on airplanes (Doglow 2012) and resulted in airport shutdown (AFP 2010).

It is apparent that analyzing the cost-effectiveness of efforts to heighten border security is meaningful for decreasing the attack risk. Considering the defender's strategies to determine which border may be most vulnerable from the terrorists' perspective (Tafoya 2011), a strategic game situation needs to be added to the cost-effectiveness analysis stemming from cyberattacks. While an attack planned by terrorists usually starts with complex strategic behaviors conducted by terrorists, at the same time, a defensive entity also needs to consider other complicated processes to prevent catastrophic results if such an attack would occur. Another point is to understand geographical and interindustrial spillover effects. Even if cyberterrorists would successfully invade a major airport system, this can cause a problem in controlling all airplane schedules and may affect not only the region in which the airport is located but also the other domestic and international airports connected to that airport. Spillover effects associated with an airport shutdown will be spread out throughout the domestic and international economies.

How can we effectively improve defensive scientific and technological instruments that address terrorists' strategies? Terrorists also have become intelligent along with modern technology innovations. Also, how can we advance an analysis tool that considers both economic damages and the probability of attack risk effectively and simultaneously? The answers involve a clearer understanding of strategic behaviors in the cyberattack procedure as well as the economic costs stemming from cyberterror events. The costs estimated should include additional losses that are extended from the direct losses related with airport system disruptions. These extended costs need to consider interindustrial connections of airport and aviation systems and negative trade effects extended to other regional losses.

In the future, it is highly expected that more air, water, and ground network systems will be affected by cyberhackers. Because of highly and increasingly ease of access to the Internet, these cyberattacks will be able to target various critical domestic and international infrastructures simultaneously. Therefore, to protect these simultaneous attacks, integrated collaborations between local governments and between the countries should be involved against these attacks.

3 The Competitive and Cooperative Game Process

Competitive game theory primarily sets strategic interactions between cyberhackers and governments (Sandler and Arce 2003). Terrorists are assumed to be intelligent and adaptive. They are able to determine where airport security measures are relatively vulnerable, utilizing various illegal network channels for transporting money, weapons, personnel, and so on. As a defender, governments would respond to potential terrorist attacks, developing potential defensive decisions on the allocation of the amount of resources that are needed to prevent any possible attack (Zhuang and Bier 2007). Based on the integrated game-theoretic interregional economic model of South Korea, Korean national policy-makers can benefit from various suggestions on which airports and/or aviation information systems primarily needed to be considered for the increased protection. The Korea Customs Service can be significantly enhanced from competitive, simulated probabilistic economic damages that are also used for the Ministry of Public Safety and Security, the Ministry of National Defense, and other agencies involved in the Korean border security.

The game-theoretic view of terrorist attack is indeed a complicated but competitive situation; the game view is not fully addressing either how local security agencies collectively cooperate to protect various borders of Korea or how national governments internationally cooperate in protecting many internationally important infrastructures from a potential attack. Some studies conducted by Frey and Luechinger (2003, 2004), Perrow (2006), and Keohane and Zeckhauser (2003) suggested ways to deter possible attackers; it is still rare to find a study on how strategic allocation of defensive resources would improve national security via in a way of reducing expected costs of a potential attack. Ideally, combining a game-

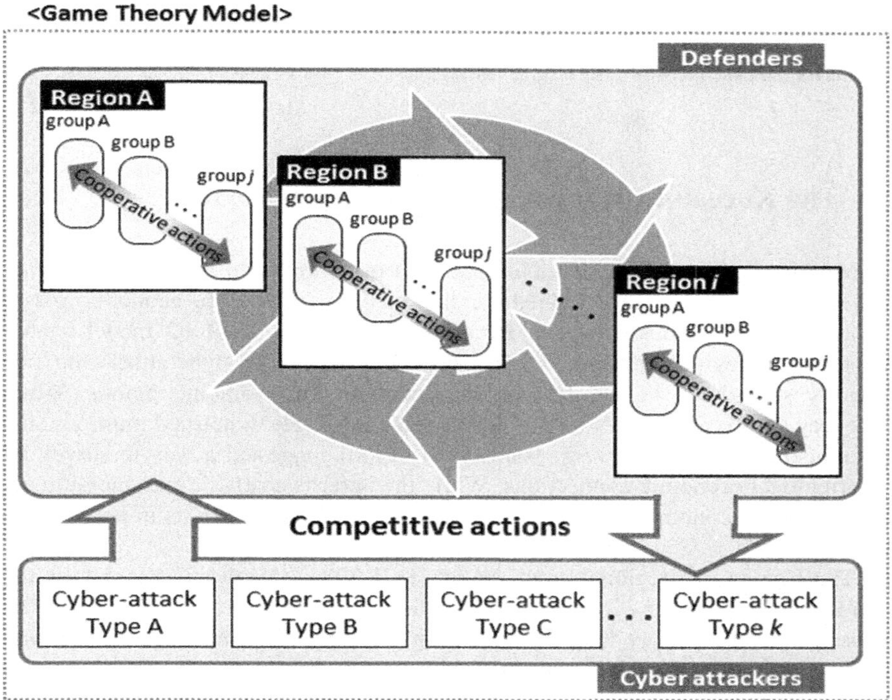

Fig. 1 Complex behavioral actions in cyberterrorism: simultaneous cooperative and competitive strategies

theoretic setting that includes both cooperative and competitive strategies with an economic impact analysis that provides economic cost estimates is one of best ways to evaluate current cyberterror security and, hence, determine the optimal future allocations of resources available for the security of infrastructure systems.

How can cooperative, collective actions be effectively applied for defense against cyberattacks in the circumstances that remove geographical and temporal boundaries? As demonstrated in Fig. 1, the integrated structure of complicated behaviors includes both cooperative and collective actions among various governments and competition actions between hackers and governments. Collaborative networking connects horizontal cooperation among local governments/groups to vertical cooperation among local and central governments/groups. This integrated, structural super-cooperation that needs optimal allocation of available resources can contribute to overcoming the threat of cyberterrorism.

While a game theory has been applied for the competitive decision processes (Benkler 2011b), a proposed public solution requires public support to be satisfied with interest groups and organizations in a society (McCain 2009). This should be publically identified and enforceable agreements. A collaborative interaction mechanism that can encourage cooperation over disastrous risk plays a more important role in the threat than the best action plan determined only via a competitive strategic behavior process and can be more effective because it is publicly agreed (Benkler 2011a; Nowak and Highfield 2011).

4 The Korean Interregional Input–Output Model

Economic models such as input–output (IO) or computable general equilibrium (CGE) models have been applied for the analysis to measure economic losses resulting from various disasters. Because one-region type of IO model cannot capture interregional effects, the interregional impacts of cyberattacks on one country should not be clarified on the economic links among regions. While Chenery (1953) and Moses (1955) developed a relatively simplified multiregional IO model (MRIO) framework, Isard (1951, 1960) suggested a way to survey all interregional economic connections. While the latter is costly, it may measure the interregional economic connections more accurately and, hence, is important as a reference data.

The Korean interregional input–output (K-IRIO) model that is released by the Bank of Korea includes 16 metropolitan cities and provinces of Korea as of 2005. Most studies that apply K-IRIO have been focusing on measuring regional and national economic benefits associated with land and infrastructure developments compared to its costs. Most impact studies that applied K-IRIO have been applied for the national level projects. As a primary tool of application to regional and national security problems, it is difficult to find a study that combined the K-IRIO with various econometric methods to quantify the costs of national security.

There is no IRIO-type model in the USA. Instead an MRIO-type model has been developed (Polenske 1980; Jack Faucett Associates 1983; Park et al. 2007). The National Interstate Economic Model (NIEMO) that is the only operational MRIO model of the USA since 1990 involves the 50 states and the District of Columbia. NIEMO has been applied to many empirical studies such as hypothetical terrorist attacks (Park et al. 2008a, b; Park 2008; Richardson et al. 2007), diverse natural disaster studies (Park et al. 2011; Park 2015), and border closure impacts of all the US borders (Gordon et al. 2009). Especially, the recent two books consecutively published by Richardson et al. (2014, 2015) include extensive empirical results stemming from various disasters using NIEMO and other economic models. Table 2 summarizes some economic impact studies to which NIEMO has been applied. The K-IRIO can be applied to these various terrorist attacks in South Korea similarly.

Table 2 Various economic impact studies using NIEMO

Nature of disruption	Targets	Type of economic impact	Total economic impacts ($M)	Base year/duration/model	Citations	Note
Explosives	LA/LB, Houston, and NY/NW	Ports shut down	23,258	2001/1 month/demand-driven NIEMO	Park et al. (2007)	Direct/indirect state-by-state impacts
Dirty bomb	LA/LB	Ports shut down	26,905	2001/1 month/supply-driven NIEMO	Park (2008)	Direct/indirect state-by-state impacts
9/11	US airports	Loss of air passengers	214,347–420,455	Sep. 2001–Aug. 2003/2 years/IMPLAN IO model	Gordon et al. (2007)	Direct/indirect/induced US impacts
Mad cow disease	US bovine animals	Cessation of foreign exports	13,681	Dec. 2003–Apr. 2005/15 months/demand-driven NIEMO	Park et al. (2007)	Direct/indirect state-by-state impacts
Explosives	13 US theme parks	Consumer losses	20,747–24,921	2004/18 months/demand-driven NIEMO	Richardson et al. (2007)	Direct/indirect state-by-state impacts
Hurricanes Katrina and Rita	US oil refinery industries	US oil price increase	1537	2005/4 months/price-sensitive supply-driven USIO	Park (2009)	Direct/indirect US impacts
Hurricanes Katrina and Rita	PADD III	Disruption of oil refinery industries	4849	2005/13 months/supply-driven FlexNIEMO	Park et al. (2016)	Direct/indirect state-by-state and month-to-month impacts
International avian influenza epidemic	Loss of air passengers, US seaports closing, loss of cross-border shopping, loss of legal and illegal labors	US border closures	1,734,075–5,408,796	2001/1 year/demand-and supply-driven USIO and NIEMO	Gordon et al. (2009)	Direct/indirect state-by-state or US impacts
Hurricanes Katrina and Rita	Louisiana customs district	Seaports shut down	44,374	Aug. 2005–Mar. 2006/8 months/demand- and supply-driven NIEMO	Park et al. (2008b)	Direct/indirect state-by-state impacts

5 The Korean Game-Theoretic Interregional Economic Model (K-GIEM)

The strategic situation of terrorists combined with K-IRIO is presented in Fig. 2, which is the distinctive feature of this study. The right panels demonstrate how the K-IRIO is connected to the simultaneous competitive and cooperative game situations in the left panels that are suggested in Fig. 1. Via the K-IRIO, the total probabilistic economic impacts will provide damage information by region, by industry, and by event type. The information will feedback to defenders and cyberattackers until the process reaches equilibrium.

As explained in Sect. 9.3, constructing the Korean game-theoretic interregional economic model (K-GIEM) requires a systematic combination of the complex game situations with a spatially disaggregate economic model. In this study, we propose an IRIO model to estimate the total costs. A general conceptual approach to calculating the framework of Fig. 2 is to multiply the success probability of a cyberattack on one (or multiple) of the main airports with the corresponding costs if the airports would be subsequently closed. By constructing a metric table of the probabilistic costs for all Korea airports considered, we can evaluate which airport may be most vulnerable in terms of cybersecurity. The dual-methodology model, combining a game-theoretic model and the K-IRIO, will generate the Korean game-theoretic interregional economic model.

The K-GIEM can estimate probabilistic economic costs when one or multiple airports would have been invaded simultaneously. In conjunction with the

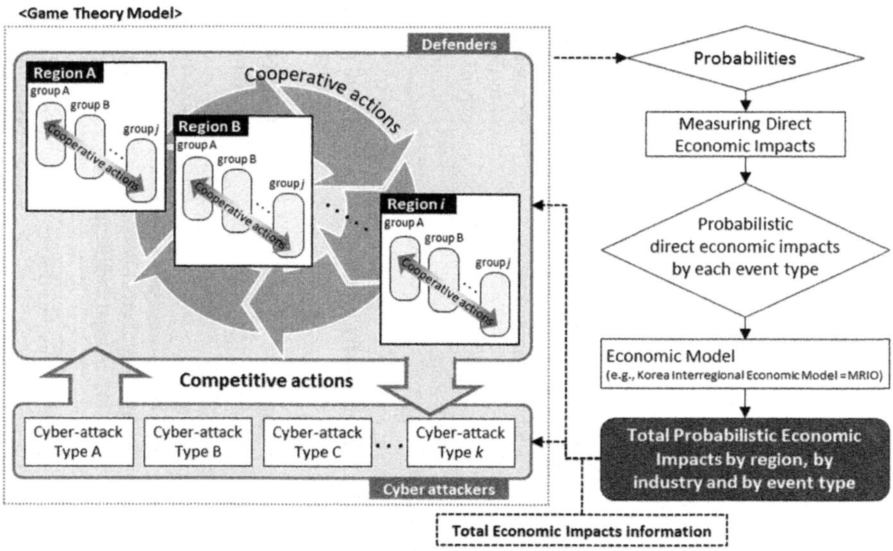

Fig. 2 A new framework to measure economic impacts using an integrated game-theoretic economic impact model

application of the game-theoretic model, the model starts analyzing the probability of each or multiple airports in Korea if attacked. It would rely on the current airport security levels and the direct cost of closure caused by successful cyberattacks. The direct cost associated with the airport closure is estimated with disruption scenarios. The product of the probability and the direct cost will then be used as input to the K-IRIO. Via the K-IRIO, the total probabilistic economic costs (by sector, by region, and by scenario) from the airport disruption will be estimated. Finally, the total probabilistic economic losses are reinformed to defenders and/or cyberattackers as new information. New economic impacts will be reestimated until reaching equilibrium. Detailed procedures are suggested as follows:

Step 1: *Modeling competitive and cooperative strategic interactions*

For one or multiple airports, the South Korea government sets a level of defensive investment allocating finance, equipment, personnel, and so on to the defensive system. A cyberattacker anticipates the government's choice and then chooses the best response level. This is the probability to launch a cyberattack against each or multiple airports. The competitive strategic interactions at this level may be applied to various forms of conflict-success functions that are used for probability estimates (Skaperdas 1996; Hausken 2004). Also, it needs to develop another form of cooperative interaction functions that set various combinations of cooperation among governments, for example, local-to-local, local-to-central, and international-governmental interactions. This type of game strategies clarifies collective defense of governments against cyberattacks. For example, simultaneous cyberattacks are plausibly made against Incheon International Airport (ICN) of South Korea and Beijing Capital International Airport (PEK) of China. In this attack, any piece of information needs to be shared in order to estimate the possible economic damages that would occur in South Korea, because most airports and routes in ICN and PEK are internationally served, connecting to each other. Furthermore, ICN is the main hub that connects to other airports in South Korea, and an attack would further disrupt other local economies in South Korea. Hence, local-to-central cooperative strategic interactions are critical for preventing cyberviolation.

Step 2: *Estimating direct costs*

To consider possible strategies for individual and cooperative defenders against cyberattackers, hypothetical and/or actual historical measures of direct costs for any successful occurrence of invasion need to be estimated. Diverse historical data can be collected from sources such as the Statistics Korea (KOSTAT), the Korea Transport Database (KTDB), and the Ministry of Employment and Labor (MOEL). These datasets should be integrated for hypothetical invasion scenarios. Based on the collected data, the direct cost of airport closure will be estimated; this will provide numerical experimental results that disclose the importance of each airport.

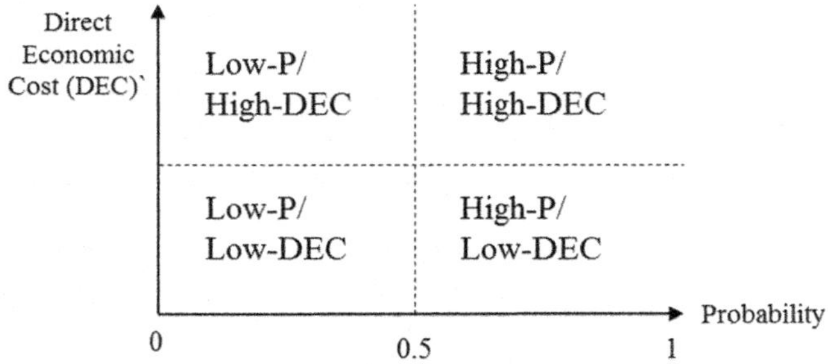

Fig. 3 Probability and direct economic cost (P/DEC) panel for cyberattack invasion

Step 3: *Estimating probabilistic direct, indirect, and total costs*

The probability of an attack success and the direct cost estimated for airport closures can be coordinated into a Probability and Direct Economic Cost (P/DEC) panel graph that is shown in Fig. 3. From the panel, we can understand where each airport fits in each panel.

It is clear that airports located in the panel of high probability (High-P) and high direct economic cost (High-DEC), whether measured collectively or individually, need to be primarily considered for the investment of Korean governments. Also, it is clear that airports in the panel of low probability (Low-P) and low direct economic cost (Low-DEC) are most trivially considered. Therefore, airports in the Low-P/High-DEC and High-P/Low-DEC panels are highly considered for optimal allocation of resources available to Korean governments. The product of the probability of a successful attack on an airport will be calculated to verify the possible direct cost resulting from the airport's closure. By aggregating the probabilistic direct costs of each airport by province in South Korea, the province's total expected direct damage can be estimated. The expected probabilistic direct costs by each province, then, are used to measure the indirect and total costs for South Korea. The estimates will be run with the K-IRIO that considers the interindustrial and interprovincial economic relations.

Step 4: *Evaluating equilibrium probabilistic impacts and vulnerable ranking metrics*

The estimated total impacts will be utilized for cyberattackers and defenders as additional information. Both attackers and defenders receive this information and decide their equilibrium strategies, resulting in an updated estimate of the total economic impacts. This process is suggested as the right and left up-pointing arrows in Fig. 2. This process will repeat until the attack probability reaches lower than a certain threshold level. The level of attack probability should be decided by Korean governments in consideration with both importance of airport security and available resources. Finally, the equilibrium total impacts will be used to construct a metric of vulnerability ranking scores that demonstrate which airports are most vulnerable from cyberattacks. The vulnerability metric may be further analyzed by industry type, by scenario, and by province of South Korea. The information provided will be used to determine which provinces and airports need primarily consideration when distributing defensive resources available in South Korea.

Step 5: *Validating K-GIEM*

Equilibrium strategies for Korean airport protection are computed using K-GIEM. To evaluate K-GIEM reliability, the computed equilibrium strategies from K-GIEM are compared to those computed from traditional game strategic models and to real datasets. The results can be used to evaluate the K-GIEM approach and may eventually help resource allocation by the Korean governments.

6 Conclusions

Aviation deterrence is crucial in fighting cyberterrorism. Aviation security against cyberinvasion is especially critical because airport networks are the heart of economic activity in that they enable rapid human and freight movement. The networks are necessary for other industrial activities. However, in many cases, cyberterrorists have been seen to be superior to governmental cybersecurity in the transportation network system (Richardson et al. 2007). As Poole (2007) suggested, this can be explained by the sizable variation in US airports, one of the three basic flaws in the Transportation Security Administration's (TSA) aviation security, which may not be clearly addressed by the TSA's centralized approach.

To effectively protect vulnerable airports from cyberterrorism, therefore, it is important to improve collaboration between local and central entities and to integrate local strategies on cyberterrorism to address the threat of simultaneous attacks on multiple Korean airports. The collaborative strategy integrates the Korean aviation security network, horizontally and vertically connecting central, local, and other nonprofit cooperative entities. This cooperative strategy should be integrated into a traditional competitive game process where such complex behavioral action strategies are set to be determined.

By combining the probability of invasion with economic costs, the K-GIEM framework is applied to quantify probabilistic economic impacts on strategic infrastructure security. It differentiates economic impacts by event location and by industry type targeted. From K-GIEM, equilibrium strategies needed for Korean airport protection can be measured. Comparing the computed equilibrium strategies from K-GIEM with those computed only by traditional game-theoretic models may provide a general guideline to evaluate optimal resource allocation by Korean governments.

One issue to be considered is resilience; many scholars agree that it refers to the defensive capacity to diminish the maximum potential impacts at any given point in time after a terrorist attack and the ability to recover as quickly as possible (Park et al. 2008a, b, 2011; McDaniels et al. 2008; Rose 2004, 2007; Adger 2000). One major way in which airport resilience from a cyberattack can take place is when, after repair, the airport resumes operation on its original schedule before airlines that use the airport persistently change to nearby airports whose services and benefits are not very different. According to Park et al. (2009b) and Rose et al. (2009), experience with the 9/11 attacks as well as with most physical disasters indicates that the economy of the USA and New York and its surrounding states had substantial resilience. Unfortunately, in contrast to factory operation to produce goods, air and water port operational services were not able to recapture the lost service values.

It is possible to measure some portion of the lost production in an economy from physical disruption by applying a fixed parameter (FEMA 1997; Rose and Lim 2002; Rose et al. 2007) or a relaxed functionalized parameter (Park et al. 2011) for each of several industrial sectors. However, most operational service sectors of infrastructure have almost no resilience. Resilience to a cyberattack may be essential to accurately estimate business interruptions that are indirectly affected by infrastructure service disruptions. However, a general framework that addresses national and international transportation network security to prevent cyberattacks still requires the combination of multidirectional, complicated game-theoretic strategies with a spatially disaggregate economic model such as K-IRIO to trace local economic activities via their connection. A further analysis of resilience is left to future research.

References

Adger, W. Neil. 2000. Social and ecological resilience: Are they related? *Progress in Human Geography* 24(3): 347–364.

AFP. 2010. *Airliners fly in face of cyber attack scares: What would happen if all those screens tracking all those flights suddenly went blank*? November 3, 2010. Available at http://news.discovery.com/tech/cyber-attack-air-traffic.html. Accessed Dec 2012.

Benkler, Y. 2011a. *The penguin and the leviathan: The Triumph of cooperation over self-interest*. New York: Crown Business.

Benkler, Y. 2011b. The unselfish gene. *Harvard Business Review* 89(7–8): 76–85.

Chenery, H.B. 1953. Regional analysis. In *The structure and growth of the Italian economy*, ed. H.B. Chenery, P.G. Clark, and V.C. Pinna, 98–139. Rome: US Mutual Security Agency.

Doglow, M. 2012. Cyberwars reach a new frontier: The airport, August 15, 2012. Available at http://www.businessweek.com/articles/2012-08-15/cyber-wars-reach-a-new-frontier-the-airport. Aaccessed Dec 2012.

Federal Emergency Management Agency (FEMA). 1997. HAZUS®[99] earthquake loss estimation methodology: User's Manual.

Frey, B.S., and S. Luechinger. 2003. How to fight terrorism: Alternatives to deterrence. *Defence and Peace Economics* 14(4): 237–249.

Frey, B.S., and S. Luechinger. 2004. Decentralization as a disincentive for terror. *European Journal of Political Economy* 20: 509–515.

Gordon, P., J.E. Moore II, J.Y. Park, and H.W. Richardson. 2007. The economic impacts of a terrorist attack on the US commercial aviation system. *Risk Analysis* 27(3): 505–512.

Gordon, P., J.E. Moore II, J.Y. Park, and H.W. Richardson. 2009. The economic costs of border closure: A state-by-state analysis. In *Global business and the terrorist threat*, ed. H.W. Richardson, P. Gordon, and J.E. Moore II, 341–374. Cheltenham: Edward Elgar.

Hausken, K. 2004. Mutual raiding of production and the emergence of exchange. *Economic Inquiry* 42(4): 572–586.
Ignelzi, L. 2012. Hackers manipulated railway computers. TSA memo says. http://www.nextgov.com/cybersecurity/2012/01/hackers-manipulated-railway-computers-tsa-memo-says/50498/. Accessed 10 June 2012.
Isard, W. 1951. Interregional and regional input–output analysis: A model of a space economy. *Review of Economics and Statistics* 33: 318–328.
Isard, W. 1960. *Methods of regional analysis: An introduction to regional science*. Cambridge, MA: The MIT Press.
Jack Faucett Associates, INC. 1983. The multiregional input–output accounts, 1977: Introduction and summary, Vol. I (Final Report). prepared for the US Department of Health and Human Services, Washington.
Kakkar, M. 2011. CBI believes cyber attack led to IGI airport's technical problems in June. September 25, 2011. Available at http://www.zdnet.com/blog/india/cbi-believes-cyber-attack-led-to-igi-airports-technical-problems-in-june/710. Accessed Dec 2012.
Keohane, N.O., and R.J. Zeckhauser. 2003. The ecology of terror defense. *The Journal of Risk and Uncertainty* 26: 201–229.
Koscher, K., A. Czeskis, F. Roesner, S. Patel, T. Kohno, S. Checkoway, D. McCoy, B. Kantor, D. Anderson, H. Shacham, and S. Savage. 2011. Experimental security analysis of a modern automobile. In *IEEE Symposium on Security and Privacy*, eds. D. Evans and G. Vigna. IEEE Computer Society, May 2010.
McCain, R.A. 2009. *Game theory and public policy*. Cheltenham: Edward Elgar.
McDaniels, T., S. Chang, D. Cole, J. Mikawoz, and H. Longstaff. 2008. Fostering resilience to extreme events within infrastructure systems: characterizing decision contexts for mitigation and adaptation. *Global Environmental Change* 18(2): 310–318.
MOLIT. 2014. 2013 Status of urban planning statistics. Ministry of Land Infrastructure and Transport (MOLIT) (in Korean).
Moses, L.N. 1955. The stability of interregional trading patterns and input–output analysis. *American Economic Review* 45: 803–832.
Nowak, M., and R. Highfield. 2011. *SuperCooperators: Altruism, evolution, and why we need each other to succeed*. New York: Free Press.
Oxford Economics. 2008. *Aviation: The real world wide web*. Available at http://www.oxfordeconomics.com/my-oxford/projects/128832. Accessed 8 July 2016.
Park, J.Y. 2008. The economic impacts of a dirty- bomb attack on the Los Angeles and Long Beach port: Applying supply-driven NIEMO (National Interstate Economic Model). *Journal of Homeland Security and Emergency Management* 5(1). doi:10.2202/1547-7355.1312.
Park, J.Y. 2009. Application of a price-sensitive supply-side input–output model to an examination of the economic impacts of hurricane Katrina and Rita disruptions of the US oil-industry, presented at the 46th annual meeting of Southern regional science association. Charleston, South Carolina, March 29–31.
Park, J.Y. 2015. Preparedness of risk society in catastrophic disasters. Monthly territory 8, KRIHS, 6–13 (in Korean).
Park, J.Y., P. Gordon, J.E. Moore II, H.W. Richardson, and L. Wang. 2007. Simulating the state-by-state effects of terrorist attacks on three major US Ports: Applying NIEMO (national interstate economic model). In *The economic costs and consequences of terrorism*, ed. H.W. Richardson, P. Gordon, and J.E. Moore II, 208–234. Cheltenham: Edward Elgar.
Park, J.Y., P. Gordon, J.E. Moore II, and H.W. Richardson. 2008a. The state-by-state economic impacts of the 2002 shutdown of the Los Angeles-long beach ports. *Growth and Change* 39(4): 548–572.
Park, J.Y., P. Gordon, J.E. Moore II, H.W. Richardson, S. Kim, and Y. Kim. 2008b. Estimating the state-by-state economic impacts of hurricane Katrina. In *Natural disaster analysis after hurricane Katrina*, ed. H.W. Richardson, P. Gordon, and J.E. Moore II, 147–186. Cheltenham: Edward Elgar.

Park, J.Y., P. Gordon, E. Jun, J.E. Moore II, and H.W. Richardson. 2009. Identifying the regional economic impacts of 9/11. *Peace Economics, Peace Science and Public Policy* 15(2). doi:10.2202/1554-8597.1162.

Park, J.Y., J. Cho, and A. Rose. 2011. Modeling a major source of economic resilience to disasters: Recapturing lost production. *Natural Hazards* 58(1): 163–182.

Park, J.Y., P. Gordon, Y.K. Kim, J.E. Moore II, and H.W. Richardson. 2016. The temporal regional economic impacts of a hurricane disaster on oil refinery operations: A flexNIEMO approach. In *Improving homeland security decisions, Forthcoming*, eds. A. Abbas, M. Tambe, and D. von Winterfeldt. New York: Cambridge University Press.

Perrow, C. 2006. Shrink the targets continued. *Spectrum* 43(9): 46–49.

Polenske, Karen R. 1980. *The US multiregional input–output accounts and model*. Lexington: DC Health.

Poole, R.W. 2007. Airport security: Time for a new model. In *The economic costs and consequences of terrorism*, ed. H.W. Richardson, P. Gordon, and J.E. Moore II, 67–97. Cheltenham: Edward Elgar.

Richardson, H.W., P. Gordon, and J.E. Moore II (eds.). 2007. *The economic costs and consequences of terrorism*. Cheltenham: Edward Elgar.

Richardson, H.W., J.Y. Park, J.E. Moore II, and Q. Pan (eds.). 2014. *National economic impact analysis of terrorist and natural disasters*. Cheltenham: Edward Elgar.

Richardson, H.W., Q. Pan, J.Y. Park, and J.E. Moore II (eds.). 2015. *Regional economic impacts of terrorist attacks, natural disasters and metropolitan policies (advances in spatial science)*. Cham: Springer.

Rose, A. 2004. Defining and measuring economic resilience to disasters. *Disaster Prevention and Management* 13(4): 307–314.

Rose, A. 2007. Economic resilience to natural and man-made disasters: Multidisciplinary origins and contextual dimensions. *Environmental Hazards* 7(4): 383–398.

Rose, A., and D. Lim. 2002. Business interruption losses from natural hazards: Conceptual and methodological issues in the case of the Northridge earthquake. *Global Environmental Change Part B: Environmental Hazards* 4(1): 1–14.

Rose, A., G. Oladosu, and S. Liao. 2007. Business interruption impacts of a terrorist attack on the electric power system of Los Angeles: Customer resilience to a total blackout. *Risk Analysis* 27 (3): 513–531.

Rose, A., G. Oladosu, B. Lee, and G.B. Asay. 2009. The economic impacts of the september 11 terrorist attacks: A computable general equilibrium analysis. *Peace Economics, Peace Science and Public Policy* 15(2). doi:10.2202/1554-8597.1161.

Sandler, T. and D.G. Arce M. 2003. Pure public goods versus commons: Benefit-cost duality. *Land Economics*, University of Wisconsin Press 79(3):355–368.

Skaperdas, S. 1996. Contest success functions. *Economic Theory* 7(2): 283–290.

Tafoya, W.L. 2011. Cyber terror. FBI law enforcement bulletin, November. Available at http://www.fbi.gov/stats-services/publications/law-enforcement-bulletin/november-2011/cyber-terror. Accessed Dec 2012.

Zhuang, J., and V. Bier. 2007. Balancing terrorism and natural disasters—Defensive strategy with endogenous attacker effort. *Operations Research* 55(5): 976–991.

Part IV
Environmental Sustainability and Policies

Effects of Urban Spatial Structure on Travel Behavior and Transportation Energy Efficiency in Korea

Kichan Nam, Brian H. S. Kim, and Up Lim

Abstract This study evaluates transportation energy efficiency with respect to various forms of urban structures. Empirical models are developed by using the population and employment of the cities in Korea. Spatial structure is measured by the degree of concentration and the clustering index and uses relative and weighted indices. A two-stage least squares model is developed because spatial structure primarily affects changes in travel distance, which consequently affect traffic volume and transportation energy consumption. Empirical results vary with city size. Small cities with deconcentrated and clustered areas are transportation energy efficient and effective in promoting clustering within a specific region and enhancing accessibility for employment and service. On the other hand, promoting concentration and declustering patterns are a more effective strategy to decentralize population, mitigate agglomeration diseconomies, and ultimately reduce transportation energy consumption in large cities. Therefore, population and employment level within a spatial distribution should be understood to adopt an optimal spatial strategy for influencing travel behavior and energy consumption.

Keywords Spatial structure • Transportation energy • Two-stage least square model • Agglomeration diseconomies

K. Nam
Korea Research Institute for Human Settlement, Anyang, Korea
e-mail: kcnam@krihs.re.kr

B.H.S. Kim (✉)
Department of Agricultural Economics and Rural Development, Interdisciplinary Program in Agricultural and Forest Meteology, and Research Institute for Agriculture and Life Sciences, Seoul National University, Seoul, Korea
e-mail: briankim66@snu.ac.kr

U. Lim
Department of Urban Planning and Engineering, Yonsei University, Seoul, Korea
e-mail: uplim@yonsei.ac.kr

E. Kim, B.H.S. Kim (eds.), *Quantitative Regional Economic and Environmental Analysis for Sustainability in Korea*, New Frontiers in Regional Science: Asian Perspectives 25, DOI 10.1007/978-981-10-0300-4_10

1 Introduction

Urban planning strategies to minimize sprawl and decentralized concentration have attracted growing interest. Traditional urban planning and development methods have been criticized for causing decentralization and increasing automobile dependency. Several studies support neo-traditional development (NTD), which involves growth management, smart growth, and new urbanism. The advantages of these approaches have been reviewed and documented in the previous studies (Brotchie 1984; Cervero 1989; Naess 1993; Newman and Kenworthy 1989; Ewing 1997). These studies claimed the advantages of compact city with high residential density with mixed land uses, where it promotes efficient public transport system and encourages walking and cycling and better opportunities for social interactions.

The debate on traditional development has intensified since 1990, particularly after the Commission of European Communities (CEC) published the "Green Paper on the Urban Environment." The paper criticizes the continual outward expansion of cities, the segregation between its residents and activities, and decentralized pattern cause of auto-dependent land development, whose harmful emissions ultimately contribute to global warming. Therefore, the CEC report also strongly supports NTD with efficient urban forms in response to growing pressures of environmentalism and sustainable urban development.

However, NTD has received criticism because of little evidence supporting the positive characterization of centralized and concentrated (compact) urban forms and the negative characterization of decentralized (sprawl) urban forms. Gordon and Richardson (1997) and Brueckner (2000) argue that less travel distance in compact urban forms increases the frequency of automobile trips that could cause increased traffic congestion and energy consumption. Consequently, the compact city theory may aggravate the congestion and energy consumption and neglect the economic advantages of decentralized city with a transportation system development and technology improvement (Gordon and Richardson 1989; Breheny 1995). Moreover, they argue that sprawl is not harmful and even has a positive effect on quality of life.

Ewing (1997) contributes with the idea of "decentralized concentration" where population or employment is not concentrated around the urban center; rather, it is concentrated in the subcenter and has a continuous development pattern. Therefore, with respect to the continual debates of urban form, exploring spatial structures is important because it provides a good foundation for understanding various activities and behavior in an urban space.

Agglomeration economies[1] significantly contribute to the understanding of spatial structure and urban growth, which are influenced by exogenous and

[1]Agglomeration economies explain the advantages of the "clustering effect" and benefit from the various activities due to urbanization, industrialization, and localization. On the other hand, agglomeration of diseconomies exists from these activities because of congestion, pollution, and low quality of life.

endogenous characteristics. Various factors in urban space (e.g., amenities, environment, accessibility, safety, economic opportunity, and health) can influence the processes of spatial structures. Richardson (1995) asserts that the processes of spatial structures can be explained by inducing positive externalities and reducing negative ones. For example, excessive concentration in a specific area can result in negative externalities, thereby changing the spatial structure and ultimately reducing agglomeration diseconomies.

Horton and Reynolds (1971) describe the stages of urban development as the process from highly unstable stage to temporary equilibrium and eventually approach to spatial equilibrium. Several factors for spatial equilibrium include quality of life (e.g., amenities and environment), economic growth, and health among others (Frumkin 2002; Sturm and Cohen 2004; Ewing et al. 2008; Royuela et al. 2010; Yang 2011). However, a controversial issue is the relationship between urban spatial structure and transportation energy because regional energy consumption with respect to spatial structure, accessibility, and transportation and the trade-off between agglomeration economies and diseconomies are important issues. Banister (1992) and Naess (1993) state that a strong relationship exists between urban form and travel activity as well as transportation energy consumption. A small-sized city may have limited access to support services and facilities, and there is a need to travel longer distances. However, a very large, centralized settlement may generate longer travel distances because the separation between homes and the urban center is larger (Owen 1986; Gordon and Richardson 1989; Banister 1992; Williams et al. 2000). Therefore, urban spatial structure factors (e.g., size, density, concentration, and clustering) that are strongly associated with transportation energy and energy-efficient spatial structures should be identified. This study performs empirical analysis by using a model to evaluate transportation energy efficiency with various forms of spatial structure.

Korea, which has high automobile ownership and use and energy consumption and whose metropolitan areas have undergone significant changes (e.g., in population, density, restoration, and infrastructure), is selected as the object of case study. In Korea, the transportation sector accounts for more than 25 % of total energy consumption, much of which is evident in the largest metropolitan areas (i.e., Seoul, Busan, Incheon, Daejeon, Gwangju, and Daegu). In the transportation sector (road, rail, marine, and air), road transportation accounts for nearly 75 % of total consumption, which explains that a large part of transportation energy is caused by inefficient urban forms in metropolitan areas in Korea. Therefore, it is particularly worthwhile to investigate transportation energy efficiency of urban areas in Korea.

2 Model Specification

Muth (1969) developed a monocentric urban model based on the assumption that all employment is located in a central business district (CBD). In the model, the number of miles a household travels depends on the distance from the household residence to the CBD, the rent gradient, and the marginal cost of travel. The models of Wheaton (1998) and Bento et al. (2005) also used the marginal time cost of travel that directly varies with population density. In their model, the marginal time cost of travel directly varies with population density, and it is inversely proportional to the land area devoted to road. Thus, the travel demand of a household is equal to the number of one-way trips to the CBD times the mean number of employed numbers of the household and depends on the road network and population distribution of the city.

In this study, the number of kilometers that a household travels and its choice of transportation mode for different types of trips depend on the spatial structure of a region. Travel demand, transportation energy consumption, and spatial structure are closely and structurally related. These relationships are specified in the spatial equilibrium-based model, which includes population and employment factors.

A household member working h hours and earning w wage per day chooses his/her residential location τ and employment based on distance d from the CBD, land consumption q, consumption goods X, and travel choice (vehicle M_v or other transportation mode M_t). Household utility can be expressed as follows:

$$U(q,\ X,\ M_v,\ M_t) \tag{1}$$

Utility is maximized by choosing an optimal location τ, quantity of goods purchased, and travel modes depending on the location. Once τ is chosen, commute mode $(Mv,\ Mt)$ and miles driven can be expressed as functions of urban form characteristics and road capacity, which are exogenous in relation to the household. The probability of travel mode choice is determined by road capacity $v(t)$, population distribution $n(t)$, travel costs (by automobile or public transit) $c(d)$, and employment distribution throughout the city.

The efficiency of spatial structure is estimated based on travel demand, especially on the number of vehicle trips per commute (*NVTPCOMM*), which is expressed as the following equation:

$$NVTPCOMM = f(PS,\ ES,\ v,\ C) \tag{2}$$

where PS is the spatial structure of population (population concentration and cluster), ES is the spatial structure of employment (employment concentration and cluster), v is the road capacity, and C is the travel cost. Gasoline consumption is determined by *NVTPCOMM*, average travel distance per vehicle (*TDPV*), and gasoline price per liter (*GASPRICE*). Therefore, gasoline consumption per commute (*GASPCOMM*) can be expressed as follows:

$$GASPCOMM = g\,(NVTPCOMM, TDPV, GASPRICE) \tag{3}$$

Following Bento et al. (2005), *TDPV* can be represented as a function of *PS* and *ES*:

$$TDPV = d\,(PS, ES) \tag{4}$$

With Eqs. (3) and (4) combined,

$$GASPCOMM = f(NVTPCOMM, d(PS, ES), GASPRICE) \tag{5}$$

Because of the distinctive relationship between agglomeration economies and diseconomies, an efficient spatial structure strategy is more reasonable than optimal spatial structure strategy depending on the size of the city. Concentration and clustering index are used for spatial structural variables related to transportation energy. The degree of centralization and concentration of the city is expected to influence travel pattern and transportation energy consumption. Interaction terms between spatial structure and population (employment) size are included in the model to account for the efficient size and spatial structure of a city. Thus, Eq. (5) is expressed as follows:

$$\begin{aligned} LNGASPCOMM = {} & \beta_1 PCON + \beta_2 PCLUST + \beta_3 ECON \\ & + \beta_4 ECLUST + \beta_5 lnP + \beta_6 (lnP) \cdot PCON + \beta_7 (lnP) \cdot PCLUST \\ & + \beta_8 (lnP) \cdot ECON + \beta_9 (lnP) \cdot ECLUST + \eta NVTPCOMM \\ & + \lambda X + \mu \end{aligned} \tag{6}$$

where LNGASPCOMM is log gasoline per commuting; PCON and PCLUST are population spatial structure (population concentration and cluster); ECON and ECLUST are employment spatial structure (employment concentration and cluster); lnP is log population; lnP*PCON, lnP*PCLUST, lnP*ECON, and lnP*ECLUST are interaction term with population and spatial structure; and NVTPCOMM is number of vehicle per commuting. Equation (6) allows us to estimate the elasticity of coefficient with respect to the population size, which is the percentage change of gasoline consumption due to the percentage change in population size (spatial structure).

2.1 Estimation of Spatial Structure

Spatial structure measurement involves many difficulties. Therefore, approaches with improved models featuring various factors (e.g., size, density, centrality, concentration, and clustering) can provide better explanations. This study employs two of the most important variables (i.e., population and employment) and the concepts used in Anas et al. (1998), Tsai (2005), and Lee (2006) to explain the global and local concentration of a regional spatial structure. Global concentration is explained as the specific-point concentration of activity at the city level, such as

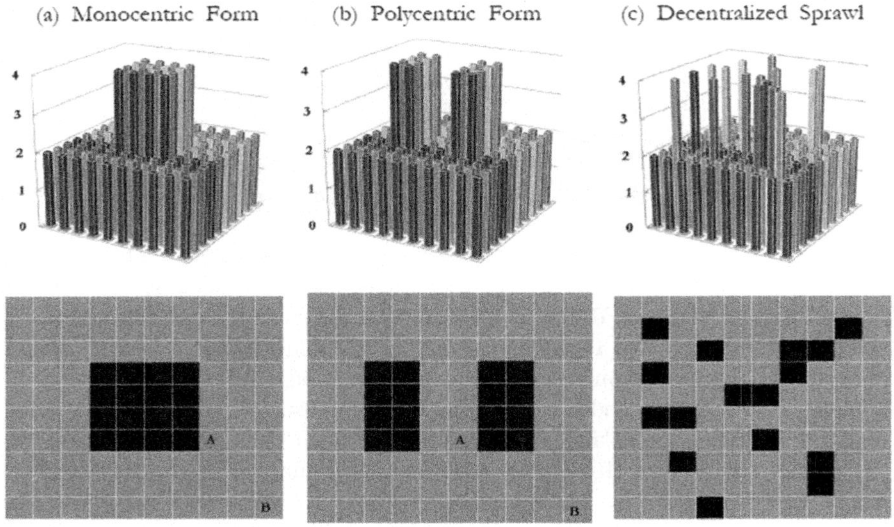

Fig. 1 Hypothesized monocentric, polycentric, and decentralized pattern

the CBD. Local concentration is explained as a disproportionate concentration of activity in different locations at the local level.

Different concentration levels are caused by the regional spatial structure, which reflects a compact, sprawl, or decentralized pattern of development based on population and employment distribution. The monocentricism or polycentricism of the region should be determined to match such patterns and measure spatial structure. However, the concepts of monocentric and polycentric forms are ambiguous. If monocentric refers to a single central area, then can this particular area have more than one center? The center refers not merely to a single center in a particular physical area but to an area with highly intense activities.

Figure 1(a–c) presents the hypothesized monocentric, polycentric, and decentralized sprawl forms, respectively, given the population size, population density, and degree of equal distribution of activities.

The shaded boxes indicate highly dense areas at the local level. The pattern shown in Fig. 1(a) is usually referred to as a monocentric city pattern. However, if the CBD is located in or around the shaded area (A or B in the figure), does this represent a monocentric or a polycentric city? That a monocentric form represents a center-concentrated pattern is a questionable assumption. Previous studies have also had difficulty in identifying a center or subcenter form of the area of study.

High clustering and unequal distribution patterns jointly characterize a monocentric city. Conversely, a polycentric city has relatively low clustering and unequal distribution. The polycentric form is the outcome of the combination of global deconcentration and local level concentration (Anas et al. 1998). If deconcentration occurs with decentralization, the area evolves into a generally dispersed form without significant subcentering (e.g., sprawl) (Lee 2006). Therefore,

Table 1 Dimensions and variables of spatial indices

Dimensions	Variables	Characteristics
Concentration	*PCONWE*	Population weighted entropy
	ECONWE	Employment weighted entropy
Clustering	*PCLUSTWE*	Population weighted Moran's I

categorizing the spatial structure of a specific area should involve the demonstration of population and employment distribution and should not be limited to the identification of a center or subcenter. Factor analysis can be applied to measure the degree of clustering and concentration of population and employment (Cutsinger et al. 2005).

Concentration and clustering indicators are measured based on the methods used by Tsai (2005). Tsai (2005) presents four indices to measure city spatial structure: size, density, concentration, and clustering. Although the scale of land area may better represent the size of the city, considering the land demand per capita and population (or employment) is a more sensible indicator. Density (or activity intensity) is the most commonly used indicator of urban form. Galster et al. (2001) describe density as the number of residential units per square mile. Ewing et al. (2002) also used residential and population density as the indicator of sprawl.

Therefore, the problems associated with confirming the characteristics of spatial structure should be resolved. Different cities may have different spatial structures, even with similar global densities; thus, population and employment distribution should be identified instead of the center. However, the density index cannot identify the city's center. It means that cities with same global densities might show different spatial structure (distribution of population and employment and its center). Thus, to identify the spatial structure, measuring the density in sublevel is more reasonable than the density in global scale.

Table 1 presents four variables representing concentration and clustering index. The Gini coefficient is generally used to measure the degree of concentration; however, it is not efficient as a relative index. Thomas (1981) reveals that the use of relative entropy (an index derived from Shannon's entropy or Theil's index that rescales the value into a range of 0–1) is a more effective method to measure the degree of concentration because it is not affected by the number of subareas. The entropy index is generally used to measure unequal distribution by comparing the population and land ratio.

Shannon's entropy index is used to measure the unequal distribution of a specific area by comparing different patterns regardless of the number of subareas. Shannon's entropy index can be defined as follows:

$$\begin{aligned}\text{Shannon Entropy} &= \sum_{i=1}^{N} Psh_i \times \log_n(1/Psh_i) \\ &= \sum_{i=1}^{N} Psh_i \times \ln(1/Psh_i)/\ln(N)\end{aligned} \tag{7}$$

where $Psh_i = P_i / \sum_{i=1}^{n} P_i$, P_i = population of subarea i, and N = number of subareas.

Tsai (2005) recommends the use of the relative entropy index based on the characteristics of Shannon's entropy index. The relative entropy index is calculated using the relative (not absolute) size of population and employment based on the characteristics of the land area (i.e., density). The relative entropy (density-based) index can be used to measure the unequal distribution of population or employment by spatial units in an urban area. Relative entropy can be defined as follows:

$$\text{Relative Entropy} = \sum_{i=1}^{n} PDEN_i \times \log\left(\frac{1}{PDEN_i}\right) / \log(N) \tag{8}$$

where $PDEN_i = DEN_i / \sum_{i=1}^{n} DEN_i$ and DEN_i is population (or employment) density at subarea i.

Moran's I, which is used to measure clustering, can be calculated in the same manner as the relative index of concentration, that is, using population and distance to other population groups:

$$Moran's\ I = \frac{N \sum_{i=1}^{N} \sum_{j=1}^{N} W_{ij} (X_i - \overline{X})(X_j - \overline{X})}{\left(\sum_{i=1}^{N} \sum_{j=1}^{N} W_{ij} \right) \sum_{i=1}^{N} (X_i - \overline{X})^2} \tag{9}$$

where N is the number of subareas i, X_i is the value of population (or employment) density at subarea i, X_j is the value of population (or employment) density at subarea j, $\overline{X}$ is the mean value of population (or employment), and W_{ij} is the spatial weight matrix corresponding to the subarea pair (i, j).

Even with more advantages and greater use of the relative index, Moran's I has limited applicability in several absolute situations (i.e., a small city with limited employment opportunities but high density). Thus, Giuliano and Small (1991) and Baumont et al. (2004) measured the spatial structure by simultaneously considering size and density. Giuliano and Small (1991) define a center as a set of continuous zones, each with a density above a particular cutoff level that, when summed, arrives at the minimum total employment. These criteria are met with a density cutoff of ten employees per acre and a minimum total employment of 10,000. Baumont et al. (2004) also suggest an alternative method, which is the exploratory spatial data analysis, and define an employment subcenter as an area with significantly higher employment and employment density than neighboring sites. Therefore, the current study considers not only the density but also the size in measuring a spatial structure. The absolute and relative indices are separately measured to calculate this index. Thus, the weighted index is calculated using the absolute and relative indices:

$$PCONWE = aPCON + bPCONRE \quad (a + b = 1) \tag{10}$$

$$PCLUSTWE = cPCLUST + dPCLUSTRE \quad (c + d = 1) \tag{11}$$

$$ECONWE = eECON + fECONRE \quad (e + f = 1) \tag{12}$$

$$ECLUSTWE = gECLUST + hECLUSTRE \quad (g + h = 1) \tag{13}$$

where *PCON, ECON, PCLUST*, and *ECLUST* are the absolute indices of population and employment concentration and clustering; *PCONWE, PCLUSTWE, ECONWE*, and *ECLUST* are the weighted indices; *PCONRE, ECONRE, PCLUSTRE*, and *ECLUSTRE* are the relative indices; and *a, b, c, d, e, f, g*, and *h* are the weights in factor analysis. This study employs weighted indices while considering the size and density.

3 Data and Variables

According to the database, 163 census tract subdivisions exist in Korea. Three census tracts were excluded from the sample because of the inappropriateness of location and condition (e.g., islands); thus, a total of 160 available census tracts are used for the analysis. Table 2 presents two levels of hierarchical administrative

Table 2 Census track for the unit analysis

		No. of census tract	No. of samples
Major cities (Si)	Seoul-si	1	1
	Busan-si	1	1
	Daegu-si	1	1
	Incheon-si	1	1
	Gwangju-si	1	1
	Daejeon-si	1	1
	Ulsan-si	1	1
	Jeju-si	2[a]	0
Provinces (Do)	Gyeonggi-do	31	31
	Gangwon-do	18	18
	Chungcheongbuk-do	11	11
	Chungcheongnam-do	15	15
	Jeollabuk-do	14	14
	Jeollanam-do	22	22
	Gyeongsangbuk-do	22 + 1[a]	22
	Gyeongsangnam-do	20	20
Sum		163	160

[a]Samples are excluded because of its location characteristic (island)

Table 3 Variables, definition, and data sources

	Variables	Definition	Data sources
Dependent	*LNGASPCOMM*	Log gasoline consumption per commuting	KNOC
Spatial structure	*LNP*	Log population size	KOSIS
	LNE	Log employment size	KOSIS
	PCONWE	Population concentration (size and density based)	Estimated value
	PCLUSTWE	Population cluster (size and density based)	Estimated value
	ECONWE	Employment concentration (size and density based)	Estimated value
	ECLUSTWE	Employment cluster (size and density based)	Estimated value
Instrumented	*NVTPCOMM*	Number of vehicle trips per commuting	KOSIS and calculated
	LNROADD	Log road density	MLTM
Instrument	*LNHHSIZE*	Log household size	KOSIS
	LNCAROWNPC	Log car ownership per capita	MLTM
Control	*LNGASPRICE*	Log gasoline price per liter	KNOC

units in Korea. First level consists of major cities and provinces called "si-do" and second level consists of small cities and counties called "si-gun."

The variables and data sources are shown in Table 3. A number of available data sources are used to produce a database for analysis. One main data source is the Korean Statistical Information Service (KOSIS), which provides census data on population demographics and housing and is one of the largest sources of census data in Korea (the census is conducted every 5 years). The data of year 2010 is used for the variables of population size, employment size, number of vehicle trips per commute, and household size (LNP, LNE, NVPCOMM, LNHHSIZE). Gasoline price and consumption data of 2010 are used, which are provided by Korea National Oil Corporation (KNOC). They are based on the gasoline sales of gasoline stations to represent transportation energy consumption. Land use data are based on the 2010 annual statistical yearbook of the Ministry of Land, Transport, and Maritime Affairs (MLTM). This statistical yearbook provides data on district divisions, population demographics and households, land size for each district, car ownership, and transportation.

Descriptive statistics of the variables are presented in Table 4. Four spatial structure variables are used, namely, *PCONWE*, *PCLUSTWE*, *ECONWE*, and *ECLUSTWE* (weighted indices). Correlation matrix is also presented in Table 5.

The correlation matrix in Table 5 shows a negative relationship between population (and employment) and PCONWE and ECONWE (−0.252 and −0.290, respectively). This negative and low correlation relationship means that larger urban size tends to have a relatively even distribution of population and

Table 4 Descriptive statistics

	Variables	Min.	Q1	Mean	Q3	Max.	Std. Dev.
Dependent variables							
Gasoline consumption	*LNGASPCOMM*	4.786	6.607	6.939	7.246	8.015	0.49
Independent variables							
Spatial structure	*LNP*	9.940	10.854	11.637	12.260	16.108	1.099
	PCONWE	0.018	0.083	0.139	0.186	0.372	0.072
	PCLUSTWE	−0.708	−0.125	0.043	0.233	0.522	0.248
	ECONWE	0.032	0.152	0.224	0.295	0.551	0.098
	ECLUSTWE	−0.790	−0.139	0.030	0.236	0.723	0.264
Instrumented	*NVTPCOMM*	0.022	0.165	0.241	0.312	0.448	0.090
Instrument	*LNROADD*	−5.531	−4.134	−3.725	−3.462	−1.689	0.615
	LNHHSIZE	4.266	4.393	4.444	4.490	4.755	0.076
	LNCAROWNPC	−1.766	−1.428	−1.340	−1.251	−0.809	0.143
Control	*LNGASPRICE*	7.111	7.122	7.128	7.131	7.142	0.007

Table 5 Correlation matrix: weighted indices

	Population			Employment		
	P	*PCONWE*	*PCLUSTWE*	*E*	*ECONWE*	*ECLUSTWE*
P	1					
PCONWE	−0.252	1				
	***					
PCLUSTWE	0.245	0.151	1			
	***	*				
E	0.993	−0.228	0.220	1		
	***	***	***			
ECONWE	−0.290	0.665	−0.062	−0.259	1	
	***	***		***		
ECLUSTWE	0.195	0.184	0.826	0.176	−0.134	1
	**	**	***	**	*	

* $p < 0.1$; ** $p < 0.05$; *** $p < 0.01$

employment, which is likely due to agglomeration diseconomies. On the other hand, population (and employment) size has a positive relationship with PCLUSTWE and ECLUSTWE (0.245 and 0.195, respectively), which indicates higher clustering pattern with larger urban population. The table also shows the relationship between spatial structure indices. Concentration and clustering indices show a strong relationship between population and employment (0.665 between PCONWE and ECONWE and 0.826 between PCLUSTWE and ECLUSTWE). Another interesting result in the table is the positive relationship of population concentration (PCONWE) with population and employment clustering indices (PCLUSTWE and ECLUSTWE, respectively). The traditional theories of spatial structure addressed that the relationship between centrality and concentration is negative in larger urban size due to agglomeration diseconomies. However, the positive relationships in the table indicate that the existing center is stronger than the suburbs in terms of the population distribution.

4 Empirical Results

Tables 6 presents the estimation results of gasoline consumption per commute with respect to the variables. The results of Models 1–4 show the effect of spatial structure on gasoline consumption. Model 1 shows the effects of population and employment spatial structure on gasoline consumption. Interaction terms between spatial structure variables and population size are included in Models 2 and 3. All spatial structure variables and interaction terms are included in Model 4. The results of Durbin-Wu-Hausman (DWH) test[2] reveal significance in all models and indicate that the two-stage least squares are more preferable than ordinary least square (OLS). The relevance of using instrumental variables is also tested using overidentifying restriction (OIR) test.[3] The results of OIR test in Table 6 indicate the significance of using instrumental variables for all models.

Instrumental variable *NVTPCOMM* has a positive coefficient, whereas control variables *LNGASPRICE* and *LNP* have negative coefficients in all models. The effects are more significant in Model 4, which considers all interaction terms and spatial structure variables. These results indicate that a greater number of vehicle trips per commute and lower gasoline price and population size result in a higher transportation energy consumption per commute. The increase in vehicle trips and energy consumption can be explained by the increase in vehicle trip efficiency, with which more vehicles may be used to travel the same distance, and the decrease in travel distance, with which more vehicle trips may be generated.

For the effect of spatial structure on gasoline consumption per commute, Model 1 presents the effect of concentration (*PCONWE* and *ECONWE*) and clustering (*PCLUSTWE* and *ECLUSTWE*) without interaction term. However, in Models 2, 3, and 4, which involve interaction terms, the effect significantly increases for both variables. The coefficient of *LNP***PCONWE* is significant and negative, whereas the coefficient of *LNP***PCLUSTWE* is significant and positive. Thus, a larger population size and a smaller population cluster indicate lower gasoline consumption. Therefore, concentration strategy to decrease in gasoline consumption is more efficient in large cities than in small cities.

Based on the estimated results in Table 6, Table 7 provides the effect of concentration and clustering based on population size. Although Table 7 provides the significant results of Table 6, the insignificant results of employment spatial structure in Model 4 are not provided in Table 7. Moreover, given that interaction terms are not employed in Model 1, the results of each model for population and employment concentration in Table 7 are the same and do not change even with incremental population size.

The coefficient of population concentration is about −2.79 when the weighted index is used in Model 2 (Table 7) in the area with a statistical mean size of the population (average *LNP* is approximately 12, as shown in Table 4). The negative coefficient of population concentration increases with increasing population size above the mean level, which indicates that large cities with high population concentration in a specific area have the propensity to decrease their transportation

---

[2] Durbin-Wu-Hausman (DWH) test statistic is used to test 2 SLS as instrumental variable (IV) against the efficiency of OLS $\text{DWH} = (\beta_{IV} - \beta_{\text{OLS}})/\sqrt{\left(S^2_{\beta_{IV}} - S^2_{\beta_{\text{OLS}}}\right)} \sim N(0,1)$; if $|\text{DWH}| > 1.96$, then X is endogenous and IV is the preferred estimator ($p < 0.05$).

[3] OIR can test whether or not the instruments are correlated with the error term in the structural model (u1). The null hypothesis of OIR test that all instruments are uncorrelated with u1 is asymptotically distributed as a Chi-squared variable with q degrees of freedom, where q is the number of instrumental variables minus the number of endogenous variables.

Table 6 Empirical result of gasoline consumption per commute: weighted index (N = 160)

2SLS	Model 1			Model 2			Model 3			Model 4		
LNGASPCOMM	Coef.		St. Err	Coef.		St. Err	Coef.		St. Err	Coef.		St. Err
Constant	68.238		46.365	97.124	**	48.908	74.026		46.845	103.9	**	51.007
NVTPCOMM	8.422	***	1.281	9.149	***	1.371	8.694	***	1.319	9.204	***	1.395
LNGASPRICE	−8.181		6.483	−12.219	*	6.853	−9.047		6.554	−13.108	*	7.123
LNP	−0.438	***	0.073	−0.47	***	0.092	−0.416	***	0.103	−0.509	***	0.111
PCONWE	−2.177	*	1.160	18.595	**	8.542	−1.741		1.146	26.443	*	13.418
PCLUSTWE	−0.240		0.307	−7.498	***	2.427	−0.099		0.313	−10.328	**	4.044
*LNP*PCONWE*				−1.782	**	0.753				−2.486	**	1.201
*LNP*PCLUSTWE*				0.637	***	0.208				0.877	**	0.343
ECONWE	1.710	**	0.822	1.541	*	0.808	9.466	*	5.662	−5.231		8.847
ECLUSTWE	0.134		0.281	0.481		0.303	−4.333	*	2.52	3.739		4.279
*LNP*ECONWE*							−0.702		0.5	0.607		0.792
*LNP*ECLUSTWE*							0.398	*	0.223	−0.279		0.368
F	10.580	***		8.544	***		8.247	***		7.011	***	
adj-R2	0.297			0.299			0.291			0.294		
DWH chi-sq test (df = 1)	33.994	***		34.664	***		34.026	***		35.959	***	
OIR test (df = 2)	3.334			1.488			2.919			1.132		

* denotes the significance at 10 %; ** denotes the significance at 5 %; *** denotes the significance at 1 %

Table 7 The effect of concentration and clustering based on the size of population:

		Population						Employment			
		Concentration			Clustering			Concentration		Clustering	
Log Pop.	Population	Model 1	Model 2	Model 4	Model 1	Model 2	Model 4	Model 1	Model 3	Model 1	Model 3
10	22,026	−2.177	0.775	1.583	−0.240	−1.128	−1.558	1.710	2.446	0.134	−0.353
11	59,874	−2.177	−1.007	−0.903	−0.240	−0.491	−0.681	1.710	1.744	0.134	0.045
12	162,755	−2.177	−2.789	−3.389	−0.240	0.146	0.196	1.710	1.042	0.134	0.443
13	442,413	−2.177	−4.571	−5.875	−0.240	0.783	1.073	1.710	0.340	0.134	0.841
14	1,202,604	−2.177	−6.353	−8.361	−0.240	1.420	1.950	1.710	−0.362	0.134	1.239
15	3,269,017	−2.177	−8.135	−10.847	−0.240	2.057	2.827	1.710	−1.064	0.134	1.637
16	8,886,111	−2.177	−9.917	−13.333	−0.240	2.694	3.704	1.710	−1.766	0.134	2.035
16.02	9,047,509	−2.177	−9.949	−13.378	−0.240	2.705	3.720	1.710	−1.779	0.134	2.042

energy consumption. Therefore, concentrated and intensive development should be stimulated to promote higher efficiency in transportation energy consumption.

Similarly, Models 2 and 4 show the effects of population clustering on gasoline consumption. The effect of population clustering is different from that of population concentration because the coefficient of population concentration is positive in average-sized cities (0.146 in Model 2; Table 7). As the city size grows, the clustering coefficient increases (1.42 in an area with 1.2 million residents in Model 2), which indicates that a more clustered area results in more gasoline consumption.

The effect of employment spatial structure is similar to that of population spatial structure. Employment concentration has a positive effect on gasoline consumption in an average-sized city (1.042 in Model 3 of Table 7), and the interaction term is negative (−0.702 in Model 3 of Table 6). Highly concentrated employment in a specific area is associated with imbalance and unequal distribution of overall activities in a city. Therefore, with a similar context of population concentration, the employment pattern in a large city should be deconcentrated to promote efficient spatial structure and energy consumption.

From an overall spatial perspective of the city, high clustering means that the specific area is spatially clustered with similar density levels. Highly dense areas cluster in a specific subarea, whereas low-density areas cluster in other subareas.

Employment and services provided in a small city (i.e., a low-density area) are relatively fewer than in a large city. Therefore, promoting clustering within a specific region is more efficient in enhancing accessibility to employment and services. However, as city size gradually increases, all services and employment clustered in such a region could cause other problems, such as congestion and long-distance travel from the outskirts of the city. Therefore, a subcenter should be formed away from the existing center as the city grows, and the overall clustering level of the city should be reduced.

Figure 2 presents a graphical illustration of these empirical results and shows the effect of spatial structure based on population and employment size. As regards the lack of scale effect without the interaction terms, the coefficients of Model 1 in Fig. 2 show a constant pattern regardless of population size. Thus, the spatial structure strategy to reduce transportation energy consumption reduces employment concentration in most cities. However, with the interaction terms involved, a declustered concentration pattern is more efficient for energy consumption in large cities, whereas a clustered deconcentration pattern is more efficient in small cities.

5 Conclusion

This paper uses an empirical model to evaluate the transportation energy efficiency of urban areas in Korea. Transportation energy efficiency is affected by various forms of spatial structure in population and employment. The empirical results vary because the effect of spatial structure varies with city size. Spatial structure is

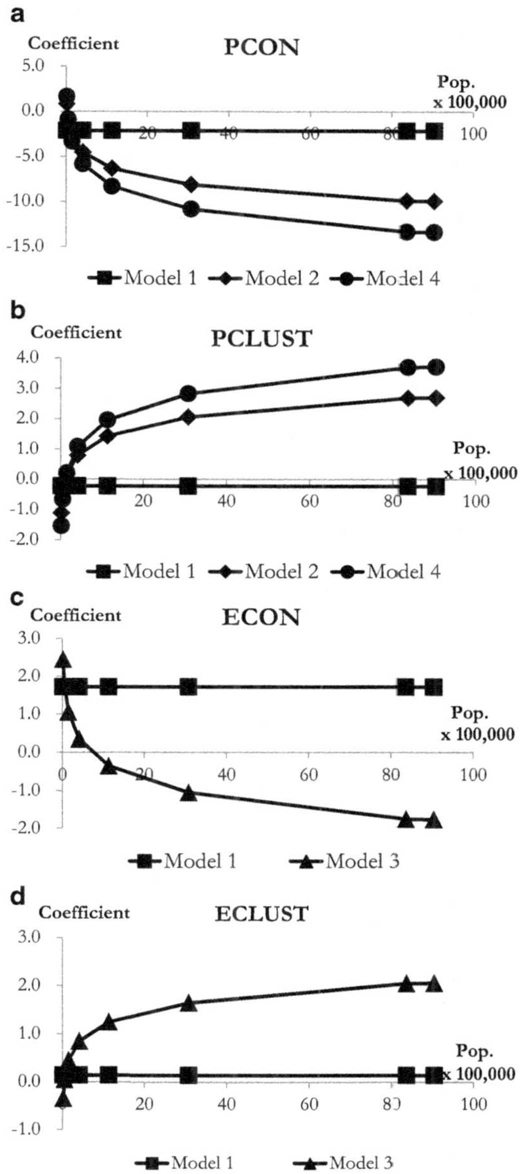

Fig. 2 Coefficient of spatial structure based on the size of population and employment: weighted index

measured by the concentration and clustering index, which uses relative and weighted indices. Each spatial unit has a different size and spatial structure; thus, the empirical test for spatial structure efficiency is conducted with interaction terms. Interaction term shows the effect of spatial structure on gasoline consumption with respect to population size. Thus, it indicates that the cities of different size need different strategies for an energy-efficient spatial structure.

For large cities, promoting concentration and declustering patterns is a more efficient strategy for minimizing gasoline consumption. Previous discussions and claims about the implementation of compact urban forms are not effective to reducing traffic volume and energy consumption because of agglomeration diseconomies (because of congestion and cross-commuting with large traffic volume in city centers). Therefore, decentralized concentration or a polycentric city type of spatial structure is necessary to decentralize population, mitigate agglomeration diseconomies, and ultimately reduce transportation energy consumption in large cities.

Small cities with a population of less than 50,000 may need to adopt a strategy different from that of large cities. Small cities with deconcentrated and clustered areas tend to be efficient in transportation energy consumption. However, if small cities adopt a large-city strategy for concentrated and declustered spatial structure, energy consumption can increase with the increased travel distance of necessary urban components. Therefore, the compact form is a more efficient strategy for small cities to reduce travel distance between the functions of urban components and to enhance transportation energy efficiency.

Thus, planners and policy makers should understand the mechanism and development of their cities to adopt and implement an optimal spatial structure strategy to influence travel behavior and energy consumption.

References

Anas, A., R. Arnott, and K.A. Small. 1998. Urban spatial structure. *Journal of Economic Literature* 36(3): 1426–1464.

Banister, D. 1992. Energy use and settlement pattern. In *Sustainable development and urban form*, ed. M. Breheny. London: Pion, Ltd.

Baumont, C., C. Ertur, and J. Le Gallo. 2004. Spatial analysis of employment and population density: the case of the agglomeration of Dijon 1999. *Geographical Analysis* 36: 146–177.

Bento, A.M., M.L. Cropper, A.M. Mobarak, and K. Vinha. 2005. The effects of urban spatial structure on travel demand in the United States. *Review of Economics and Statistics* 87: 466–478.

Breheny, M. 1995. Compact cities and transport energy consumption. *Transactions of the Institute of British Geographers* 20(1): 81–101.

Brotchie, J.F. 1984. Technological change and urban form. *Environment and Planning A* 16: 583–596.

Brueckner, J.K. 2000. Urban sprawl: Diagnosis and remedies. *International Regional Science Review* 23(2): 160–171.

Cervero, R. 1989. Jobs-housing balancing and regional mobility. *Journal of the American Planning Association* 55: 136–150.

Cutsinger, J.G., H.W. Galster, R. Hanson, and D. Towns. 2005. Verifying the multi-dimensional nature of metropolitan land use: Advancing the understanding and measurement of sprawl. *Journal of Urban Affairs* 27: 235–259.

Ewing, R. 1997. Is Los Angeles-style sprawl desirable? *Journal of the American Planning Association* 63: 107–126.

Ewing, R., R. Pendall, and D. Chen. 2002. *Measuring sprawl and its impact*. Washington, DC: Smart Growth America.

Ewing, R., T. Schmid, R. Killingsworth, A. Zlot, and S. Raudenbush. 2008. Relationship between urban sprawl and physical activity, obesity and morbidity. *Urban Ecology* 18(1): 567–582.
Frumkin, H. 2002. Urban sprawl and public health. *Public Health Reports* 117(3): 201–217.
Galster, G., R. Hanson, M.R. Ratcliffe, H. Wolman, S. Coleman, and J. Freihage. 2001. Wrestling sprawl to the ground: Defining and measuring an elusive concept. *Housing Policy Debate* 12 (4): 681–717.
Giuliano, G., and K. Small. 1991. Subcenters in the Los Angeles region. *Regional Science and Urban Economics* 21(2): 163–182.
Gordon, P., and H.W. Richardson. 1989. Gasoline consumption and cities: A reply. *Journal of American Planning Association* 55: 342–345.
Gordon, P., and H.W. Richardson. 1997. Are compact cities a desirable planning goal? *Journal of the American Planning Association* 63: 95–106.
Horton, F.E., and D.R. Reynolds. 1971. Effects of urban spatial structure on individual behavior. *Economic Geography* 47(1): 36–48.
Korea national oil corporation (KNOC) Opinet (www.opinet.co.kr) and Petronet (www.petronet.co.kr) database.
Korean statistical information service (KOSIS) database (www.kosis.kr).
Lee B. 2006. Urban spatial structure, commuting, and growth in US metropolitan areas. Unpublished doctoral dissertation, University of Southern California.
Muth, R.F. 1969. *Cities and housing*. Chicago: The University of Chicago Press.
Naess, P. 1993. Transportation energy in Swedish towns and regions. *Scandinavian Housing and Planning Research* 10(4): 187–206.
Newman, P., and J. Kenworthy. 1989. Gasoline consumption and cities. *Journal of the American Planning Association* 55(1): 24–37.
Owens, S.E., 1986. *Energy, planning, and urban form*. London: Pion Ltd.
Richardson, H.W. 1995. Economies and diseconomies of agglomeration. In *Urban agglomeration and economic growth*, ed. H. Giersch. Berlin: Springer.
Royuela, V., R. Moreno, and E. Vayá. 2010. Influence of quality-of-life on urban growth: a case study of Barcelona, Spain. *Regional Studies* 44(5): 551–567.
Sturm, R., and D. Cohen. 2004. Suburban sprawl and physical and mental health. *Public Health* 118(7): 488–496.
Thomas, R.W. 1981. *Information statistics in geography*. Norwich: Geo Abstracts.
Tsai, Y.H. 2005. Quantifying urban form: Compactness versus sprawl. *Urban Studies* 42: 141–161.
Wheaton, W.C. 1998. Land use and density in cities with congestion. *Journal of Urban Economics* 43: 258–272.
Williams, K., E. Burton, and M. Jenks. 2000. *Achieving sustainable urban form*. London: E & FN Spon.
Yang, S.W. 2011. The compact city and networked city in Korea. *International Journal of Urban Science* 7: 193–203.

Need for Coordination between Greenhouse Gas and Pollution Abatement Regulations: China's Case and Its Implications for Korea

Kyung-Min Nam

Abstract In this chapter, I briefly introduce my previous research on the carbon co-benefits of pollution abatement in China and then discuss what key lessons Korea can learn from it. A main finding is that strong cross effects exist between air pollution and carbon mitigation in China. In particular, China could even over-achieve its official CO_2 intensity targets, in terms of carbon reductions, by simply meeting the existing SO_2 and NO_x reduction goals. Accordingly, the CO_2 intensity targets are not binding and generate unnecessary compliance costs. This result conveys several policy implications for Korea. First, local pollution abatement, given its strong cross effects, may be considered as a carbon mitigation strategy. However, taking full advantage of the cross effects—meeting emission reduction targets at minimal costs—requires a close coordination between air pollution and carbon regulations. Finally, clear and consistent long-term reduction goals and associated policy incentives are necessary to promote economy-wide, forward-looking technology adoption and thus to avoid the potential lock-in effect in energy supply.

Keywords CGE model • Co-benefits • Air pollution • Carbon mitigation • China • Korea

1 Introduction

At the 2009 United Nations Climate Change Conference, the Korean government officially announced that by 2020 it would reduce national greenhouse gas (GHG) emissions from the business-as-usual (BAU) level by 30 %. To achieve this goal, Korea has introduced a new GHG control system, consisting of a Target Management Scheme (TMS) and an Emissions Trading Scheme (ETS). The TMS imposes emission caps on a group of firms and installations, which either emit a great deal of

K.-M. Nam (✉)
Department of Urban Planning and Design, The University of Hong Kong, Pokfulam Road, Hong Kong
e-mail: kmnam@hku.hk

E. Kim, B.H.S. Kim (eds.), *Quantitative Regional Economic and Environmental Analysis for Sustainability in Korea*, New Frontiers in Regional Science: Asian Perspectives 25, DOI 10.1007/978-981-10-0300-4_11

GHGs or consume a large amount of energy.[1] Of those firms which are subject to the TMS or operate TMS-regulated installations, 526 have been allocated carbon permits and are participating in the ETS during the first phase (2015–2017).[2]

In addition, Korea has a long history of air pollution control. The first nationwide air quality standards were prepared for sulfur dioxide (SO_2) in 1978. Since then, Korea has extended the scope of its regulations to include other key conventional air pollutants, such as nitrogen dioxide (NO_2), carbon monoxide (CO), particulate matter (PM_{10}), and volatile organic compounds (VOCs). At present, Korea's air quality regulations target motorized vehicles and pollution-intensive industrial facilities, placing restrictions on end-of-pipe concentrations. In addition to this nationwide regulation, the Capital Region,[3] where air pollution is more serious than in other parts of Korea, has adopted further proactive control measures, implementing strict region- and firm-specific emission caps since 2005.

Then, how well coordinated are these two independent regulatory systems, each aiming at GHG reduction or air quality control? GHGs and conventional air pollutants are often cogenerated from the same sources, such as combustion of fossil fuels (Agee et al. 2014; Shindell et al. 2011). Accordingly, a regulation targeting GHG reductions automatically affects air pollutant emissions and vice versa (Bollen et al. 2009; Morgenstern et al. 2004; Xu and Masui 2009). Neglect of such unintended air pollution abatement from GHG reductions (i.e., air quality co-benefits of GHG reductions) or of ancillary GHG reductions from pollution abatement (i.e., climate co-benefits of pollution abatement) may entail overly large policy compliance costs, leading to a suboptimal market outcome (Nam et al. 2014). That is, achieving GHG and pollutant reduction targets at minimal costs would require a careful consideration of synergistic effects between GHG and pollution regulations. However, there is little evidence that such synergistic effects have been reflected in Korea's current GHG reduction goals, leaving room for further discussion (Chae and Park 2011).

In this chapter, I discuss China's case and draw key implications for Korea, focusing on the inseparable nature of GHG and pollution abatement. For my discussion, I first take a brief look at key facts relating to carbon emissions and air pollution in Korea and explore why the co-benefits argument may also be relevant to Korea's context. Then, I briefly introduce China's official pollution and GHG regulation targets and their potential synergistic effects, which are summarized from my earlier coauthored study (Nam et al. 2013). Finally, I discuss what lessons Korea can learn from China's case.

[1] At present, the TMS targets firms exceeding 50 $ktCO_2e$, in terms of annual GHG emissions, or 200 TJ, in terms of annual energy consumption, and the installations exceeding 15 $ktCO_2e$ or 80 TJ (Government of Korea, 2014).

[2] These 526 ETS participants either emitted an annual average of $\geq$125 $ktCO_2e$ of GHGs between 2011 and 2013 or operated an installation emitting $\geq$25 $ktCO_2e$ during the same period (Government of Korea, 2014).

[3] The Capital Region includes the following three province-level municipalities: Seoul, Incheon, and Gyeonggi.

2 Where Korea Stands

2.1 *GHG Emissions*

Korea is one of the world's major GHG emitters, in both gross and per capita terms. In 2011, Korea emitted 624 Mt of carbon dioxide (CO_2), which accounted for 1.7 % of the global emissions, and was the world's eighth largest emitter (Fig. 1). Korea's carbon emissions may be considered even larger if seen in per capita terms. In 2011, for example, Korea's per capita CO_2 emissions were 11.8 metric tons, exceeding the global average by a factor of >2. About 85 % of the total CO_2 emissions are from fuel combustion, and over half the fuel-related emissions are from solid sources (mainly coal), followed by liquid fuels. With growing demand for energy, particularly its solid component, Korea's CO_2 emissions have shown a continuously increasing trend.

The Korean government has taken this situation seriously, expressing its strong will to join the global effort to reduce GHG emissions. In the Copenhagen Accord, Korea stated its GHG mitigation target for 2020 as 30 % below the BAU level, which corresponds to the upper limit of the range recommended by the Intergovernmental Panel on Climate Change (IPCC). At present, this target is legally binding, since it is specified in the Framework Act on Low Carbon, Green Growth, enacted in January 2010. Sector-specific reduction targets and implementation details have been announced through several follow-up plans and policies, such as the Roadmap to Meeting the National GHG Reduction Goals. As shown in Table 1, the industry and power/heat-generation sectors account for >60 % of the total reductions, and the buildings and transport sectors are responsible for the majority of the remaining reductions. Primary channels of meeting the targets

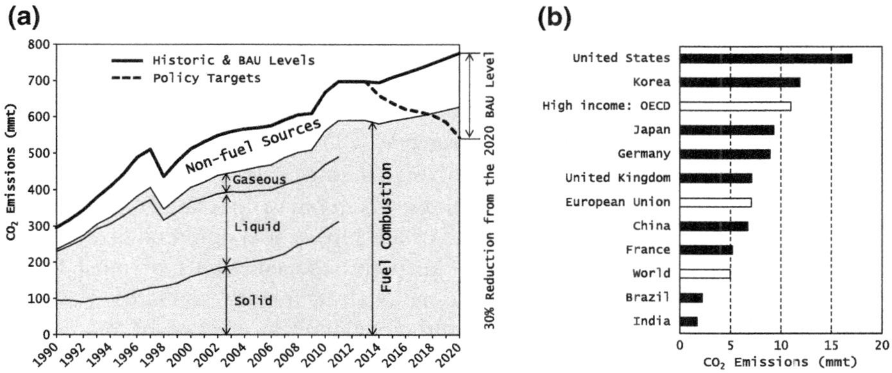

Fig. 1 Korea's CO_2 emission trend and per capita emissions: (**a**) CO_2 emissions by source, (**b**) CO_2 emissions per capita, 2011. Note: CO_2 emissions presented here exclude those discharged from land use, land use change, and forestry (Source: Created by the author from the Korea Energy Statistics Information System (KESIS) and the World Development Indicator (WDI) databases)

Table 1 Korea's 2020 CO_2 reduction targets by sector

	Emissions ($MtCO_2e$)		Reduction		
Sector	BAU	Target	($MtCO_2e$)	% of BAU	% of total
Total	776.1	543.0	233.1	30.0	100.0
Industry	439.0	357.7	81.3	18.5	34.9
Buildings	167.6	122.6	45.0	26.8	19.3
Transport	99.6	65.4	34.2	34.3	14.7
Agriculture, forestry, fishery	28.5	27.0	1.5	5.3	0.6
Waste	13.8	12.1	1.7	12.3	0.7
Others	17.9	13.4	4.5	25.1	1.9
Transformation*		−64.9	64.9		27.8

Source: Modified from the Government of Korea (2014), p. 11
Note: *Here, "transformation" refers to emission reduction attained by the power and heat generation sector through switching toward less carbon-intensive fuels and expanding adoption of renewable or carbon-free energy technologies

include fuel switching toward less carbon-intensive sources, cleaner energy technology adoption, and improved energy intensity, which will be encouraged through a mixture of command-and-control regulations (e.g., renewable portfolio standards) and market-based policy incentives (e.g., ETS).

2.2 *Air Pollution*

At present, Korea implements nationwide air quality standards for seven pollutants (Table 2). The SO_2 standards prepared in 1978 are the forerunner of Korea's air quality control and are followed by those for NO_2, ozone (O_3), CO, and fine particles formulated in 1983. Fine particles were initially regulated in terms of total suspended particulate (TSP) concentrations, but the PM_{10} and $PM_{2.5}$ standards, set in 1993 and 2012, respectively, have completely replaced the TSP standards since 2001. Lead (Pb) and benzene (C_6H_6) concentrations have been controlled since 1991 and 2007, respectively. Industrial facilities must meet the government-set emission standards, and those which fail to do so are required either to install additional abatement equipment or are subject to a pollution tax.

Korea has successfully managed to keep the concentrations of most target pollutants below the upper limits of the national air quality standards. Between 2006 and 2014, for example, O_3 concentrations showed a range of 21–27 ppb (or 41–52 $\mu g/m^3$), in terms of annual means of 8-h daily maximum, which is substantially below the upper threshold of 60 ppb (Fig. 2). Similarly, NO_2 and SO_2 showed ranges of 24–27 ppb (or 44–50 $\mu g/m^3$) and 5–6 ppb (or 13–16 $\mu g/m^3$), respectively, which are substantially lower than the air quality standards of 30 ppb and 20 ppb. PM_{10} levels exceeded 50 $\mu g/m^3$ until 2010 but have been maintained below the threshold since 2011.

Table 2 Korea's national air quality standards, as of 2015

Pollutant	Annual means	Daily means	8-h means	1-h means
SO_2	0.02 ppm	0.05 ppm	–	0.15 ppm
CO	–	–	9 ppm	25 ppm
NO_2	0.03 ppm	0.06 ppm	–	0.10 ppm
Fine particles				
PM_{10}	50 μg/m^3	100 μg/m^3	–	–
$PM_{2.5}$	25 μg/m^3	50 μg/m^3	–	–
O_3	–	–	0.06 ppm	0.1 ppm
Pb	0.5 μg/m^3	–	–	–
C_6H_6	5 μg/m^3	–	–	–

Source: Created from the Ministry of Environment, Korea (http://eng.me.go.kr/eng/web/index.do?menuId=253&findDepth=1)

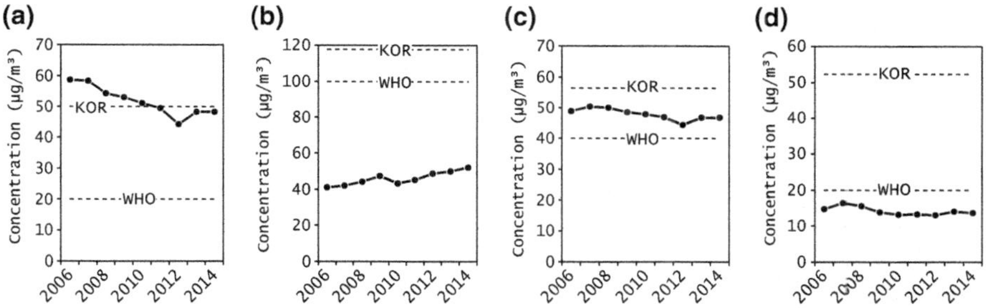

Fig. 2 Concentration trends of selected air pollutants in Korea, 2006–2014: (**a**) PM_{10}, (**b**) O_3, (**c**) NO_2, (**d**) SO_2. Note: Numbers displayed are the population-weighted average of province-specific annual mean concentrations. KOR: Korea's national air quality standards. WHO: WHO guideline levels (Source: Created from the National Institute of Environmental Research (NIER) database)

One point to be noted, however, is that the Korean air quality standards are relatively moderate compared with those implemented in the West or recommended by international organizations, such as the World Health Organization (WHO). Korea's PM_{10} and NO_2 levels are particularly higher than the WHO guideline levels (20 μg/m^3 and 40 μg/m^3, respectively), which are widely adopted to manage pollution-induced health risks within a reasonable range (WHO 2006). Accordingly, a voice demanding stricter air quality standards has been growing, given that their primary purpose involves public health concerns.

A public response to the demand is stricter antipollution regulations targeting the Capital Region, where half the Korean population resides (MOE 2010). In December 2003, the Korean government legislated the Special Act on Seoul Metropolitan Air Quality Improvement, and the first basic and implementation plans, covering the period of 2005–2014, followed in November 2005 and

Table 3 Pollution control targets for the capital region

	First plan (2005–2014)*		Second plan (2015–2024)**	
Pollutant	Emissions[a]	Concentration	Emissions[b]	Concentration
NO_x	53 % (164 kt)	22 ppb	55 % (300 kt)	20–21 ppb[e]
SO_x	39 % (27 kt)	–	44 % (50 kt)	–
PM_{10}	39 % (6 kt)[c]	40 μg/m^3	59 % (68 kt)[d]	30–37 μg/m^3[f]
$PM_{2.5}$	–	–	45 % (14 kt)[d]	20 μg/m^3
O_3	–	–	–	60–70 ppb[g]
VOCs	39 % (102 kt)	–	56 % (304 kt)	–

Source: Created from MOE (2010; 2013)
Note: *Target year is 2014; **target year of achieving given policy goals is 2024
[a]Emission reduction below the 2001 level
[b]emission reduction below the 2024 BAU level
[c]Targets exclude fugitive dust emissions
[d]targets include fugitive dust emissions
[e]21 ppb for Seoul and 20 ppb for Incheon and Gyeonggi
[f]30 μg/m^3 for Seoul, 36 μg/m^3 for Incheon, and 37 μg/m^3 for Gyeonggi
[g]60 ppb for Seoul and Incheon and 70 ppb for Gyeonggi

December 2006, respectively. The plans incorporate emission-based pollution abatement schemes into conventional concentration-centered air quality regulations, strengthening their proactive stance. Target pollutants on which emission caps are imposed include PM_{10}, nitrogen oxides (NO_x), sulfur oxides (SO_x), and VOCs, and the trade of emission permits is allowed among participating firms only for NO_x and SO_x. The 2014 reduction goals include a 53 % reduction from the 2001 level for NO_x and a 39 % reduction for the other three pollutants (Table 3). NO_x targets are more stringent than the others, as NO_x pollution is particularly serious in the Capital Region (Fig. 3). The region's PM_{10} levels are also high, but do not deviate from the national means as much as its NO_x levels do. Follow-up plans, covering the period of 2014–2024, adopt more stringent pollution abatement and air quality targets and add O_3 and $PM_{2.5}$ to the list of criteria pollutants (MOE 2013).

2.3 *Primary Emission Sources*

In Korea, as is the case for many other countries, GHGs and conventional air pollutants largely share common sources of emissions. In 2011, for example, fuel combustion accounted for 84 % of Korea's anthropogenic CO_2 emissions and was also the main culprit responsible for the nation's air pollution, contributing to 82 % of SO_x, 95 % of NO_x, and 94 % of PM_{10} emissions (Fig. 4). This overlap may speak for potentially large cross effects between pollution and carbon regulations.

Further sectoral decomposition clarifies that power generation, industrial combustion, and road transportation are three major common sources of carbon and air pollutant emissions. In the case of CO_2, 87 % of the emissions were from stationary sources, such as power and heat generation (36 %), industrial combustion (18 %),

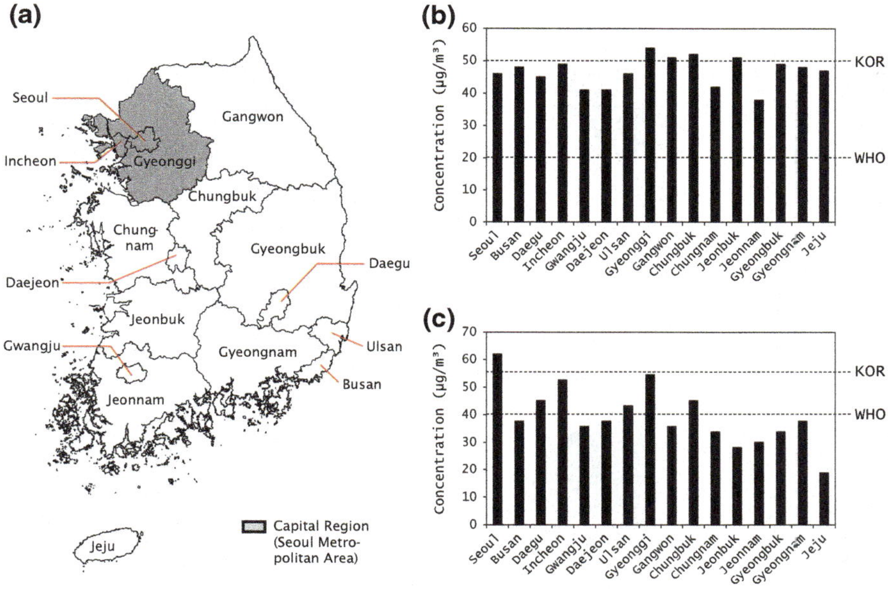

Fig. 3 PM_{10} and NO_2 concentrations in Korea, by province, 2006–2014: (**a**) map of Korea, (**b**) PM_{10}, 2014, (**c**) NO_2, 2014. Note: KOR: Korea's national air quality standards; WHO: WHO guideline levels (Source: Created from the NIER database)

and residential combustion (18 %), and most of the remainder was from road transportation (11 %). Stationary sources also accounted for ≥79 % of SO_x and PM_{10} emissions: industrial combustion (25 %), industrial process (25 %), and electricity production (19 %) were almost equally important stationary sources for SO_x, while industrial combustion (68 %) was the single most important emission source of PM_{10}. When it comes to mobile sources, only non-road transportation (16 %) was crucial for SO_x, while both road (10 %) and non-road transportation (11 %) were comparably important for PM_{10}. In contrast to CO_2 and the two pollutants, over half the NO_x, emissions were from mobile sources, making road transport (31 %) and non-road transport (21 %) the two most important sectoral contributors. Among stationary sources, industrial combustion (17 %) and power generation (16 %) accounted for a large share of NO_x emissions.

Despite such overlapped emission sources, Korea has implemented the two regulatory schemes (i.e., air pollution and GHG control) in a rather independent manner (Chae 2010), a situation that may result in substantial efficiency loss. In the following section, I will briefly introduce the Chinese case and demonstrate that the cost of poor coordination between the policies can be substantially large.

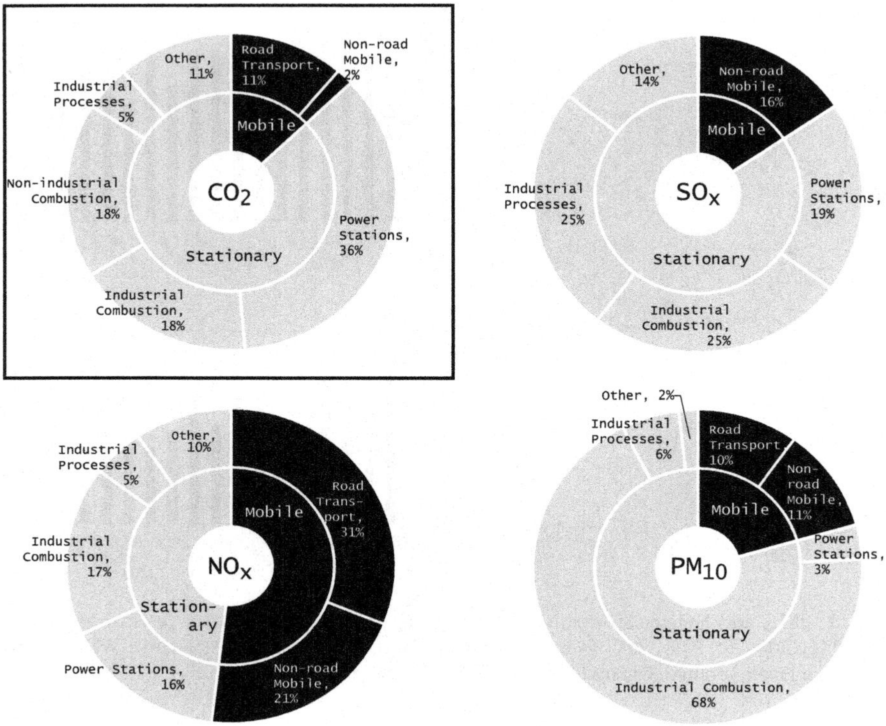

Fig. 4 Decomposition of CO_2 and selected air pollutant emissions in Korea by source, 2011 (Source: Created from the Organisation for Economic Co-operation and Development (OECD) database)

3 Carbon Co-benefits of NO_x and SO_2 Abatement in China

3.1 Key Facts on Emissions

Air pollution in China is notorious for its magnitude. As of 2012, for example, PM_{10} concentrations in 31 Chinese major cities exceeded the WHO-recommended level by a factor of 1.7–7.3, and only one of them—Haikou—met the national class 1 standards (Fig. 5). A consequence of such excess air pollution is an increased threat to public health. Health damage associated with excess PM_{10} pollution alone is estimated to be as large as 4–9 % of China's historic gross domestic product, causing a substantial loss of economy-wide welfare (Matus et al. 2012; Nielsen and Ho 2013; World Bank and SEPA 2007). With a growing concern about the situation, China has implemented the National Ambient Air Quality Standards (NAAQS) since 1982 (Wang and Hao 2012). At present, the NAAQS covers ten pollutants, including SO_2, NO_x, PM_{10}, and O_3, and its current implementation in prefecture-level cities will be extended to the entire nation from 2016 (MEP 2012).

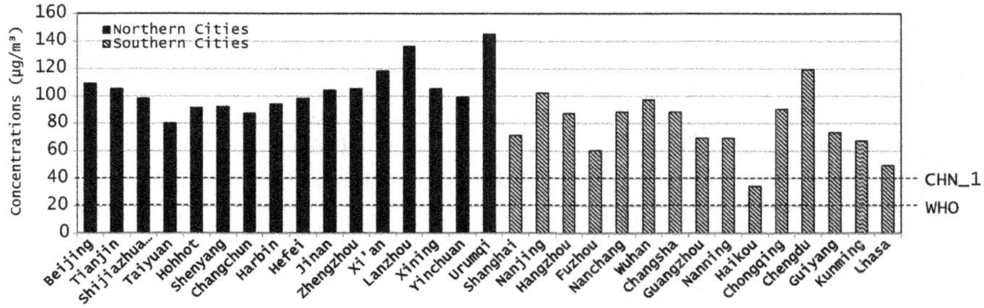

Fig. 5 Annual mean PM_{10} concentrations in selected Chinese major cities, 2012. Note: CHN_1: class 1 standards included in NAAQS; WHO: WHO guideline levels (Source: Created from the China Statistical Yearbook, 2013)

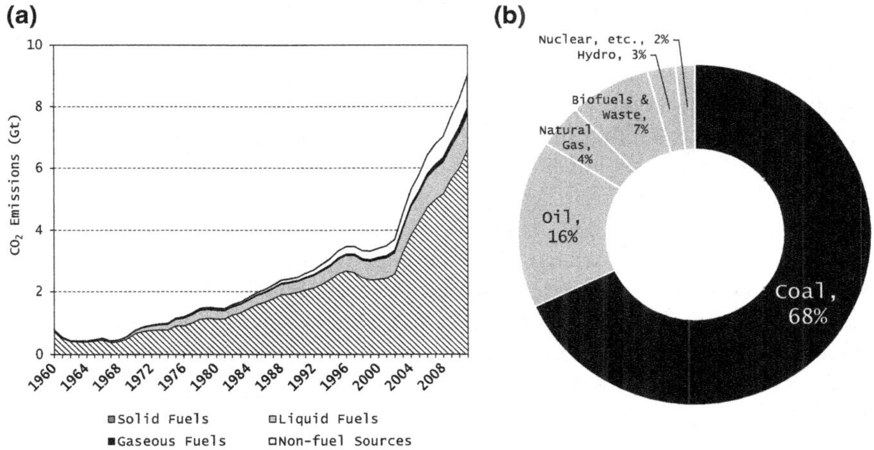

Fig. 6 Historic CO_2 emissions in China and their decomposition by fuel source: (**a**) CO_2 emissions by source, 1960–2011, (**b**) decomposition of primary energy supply, 2011 (Source: Created from the WDI and the International Energy Agency (IEA) databases)

China is currently the world's largest CO_2 emitter, and it has presented a rapidly increasing emission trend during the last decade (Fig. 6). The soaring carbon emission trend in China is due mainly to the interplay of rapid economic growth, low energy efficiency, and high dependence on coal. Rapid growth necessarily drives up energy demand, and it is more so in China, given that its economy has a highly energy-intensive structure. In 2012, for example, China's energy efficiency was below half the EU level and 69 % of the US level.[4] What makes the situation even worse is high dependence on coal, which has higher emission factors than

[4] Energy efficiency is defined as value added per unit energy use, and the numbers given here are computed from the WDI database.

other fossil fuels. In 2011, for example, over two-thirds of China's energy demand was fulfilled by coal, and solid fuels account for 69–96 % of China's historic carbon emissions for the last half century. For this reason, the Chinese government has implemented energy and carbon intensity reduction goals, targeting the cutback of coal use by industrial and energy sectors (Liu et al. 2013).

Against this background, the 12th Five-Year Plan (FYP12)—the most recent roadmap of the Chinese economy covering the period between 2011 and 2015—presents binding reduction targets for two pollutants (NO_x and SO_2) and CO_2. The NO_x and SO_2 targets are given as a 10 % and an 8 % *emission* reduction from the 2010 levels, respectively, and the CO_2 target is given as a 17 % *intensity* reduction from the 2010 level.

3.2 Motivation for Co-benefit Analysis

Despite the legally binding pollution and carbon abatement targets, their coordination may be in question. As is the case for Korea, the overlapping of sectoral emission sources is striking in China. In 2010, for example, 66–82 % of the NO_x, SO_2, and CO_2 emissions in China were from the power generation and industrial sectors, suggesting potential synergy between pollution and carbon abatement regulations: the NO_x and SO_2 targets will likely place strong constraints on the energy use and mix associated with these two sectors, which may lead to a substantial unintended CO_2 reduction. However, the FYP12 CO_2 intensity targets seem to be too moderate to actualize such potential synergy, given that carbon emissions in absolute terms will still grow even when the targets are met, due to China's fast economic growth. Our analysis introduced in this section is motivated to test this hypothesis (Fig. 7).

3.3 Method[5]

For hypothesis testing, my colleagues and I extend the fifth version of the MIT Emissions Prediction and Policy Analysis (EPPA) model, which is a recursive dynamic, multiregional computable general equilibrium (CGE) model built on the Global Trade Analysis Project version 7 dataset (Chen et al. 2015; Paltsev et al. 2005). Our effort focuses on explicit representation of pollution abatement structure within a CGE framework. For fuel-related emissions, we first create a fuel pollution nest, where a unit of fuel combustion (X_F) is associated with a unit of air pollution (X_P) in a fixed proportion (i.e., Leontief) (Fig. 8a). Then, we create a sub-nest, and let pollution, once generated, be either emitted (X_E) or abated before

[5] For further details on the method introduced in this section, refer to Nam et al. (2013) and (2014).

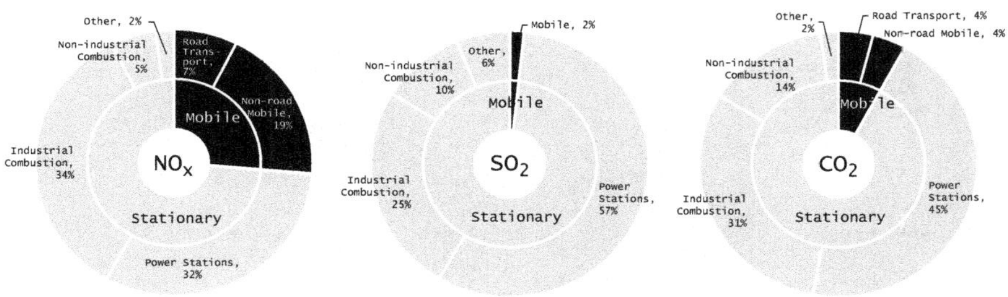

Fig. 7 Decomposition of NO_x, SO_2, and CO_2 emissions in China by source, 2010 (Source: Created from the IEA and GAINS-China Databases)

emission (X_A) through end-of-pipe control measures. Here, X_P is modeled as a constant elasticity of substitution (CES) production function of X_E and X_A with elasticity σ_{Fuel}.We estimate σ_{Fuel} by fuel, sector, and pollutant, using detailed engineering data from the Greenhouse Gas and Air Pollution Interactions and Synergies (GAINS) database.

An optimal emission abatement share is endogenously determined within the model, considering the stringency of pollution caps and associated abatement costs. In the absence of pollution control, the cost of emissions is zero, so all pollution generated will be emitted. However, when pollution is regulated, each unit of pollutant emissions carries a cost, and the cost schedule follows the path of a conventional marginal abatement cost (MAC) curve. As exemplified in Fig. 9, the MAC increases slowly at the beginning, but its slope becomes steeper when the remaining abatement potential is increasingly exhausted. Near the end of the right tail of the curve, adoption of end-of-pipe technologies for additional abatement is extremely expensive, and thus firms will eventually have to reduce pollution through less use of energy, as well as fuel switching.

Nonfuel pollution is treated as a conventional production input, and its abatement requires a proportional increase of other inputs in the nest when their prices are unchanged (Fig. 8b). The substitution elasticity between pollution and other inputs is given as $\sigma_{Pollutant}$, and is also estimated by fitting GAINS data to a log-linear equation.

CO_2 emissions and fossil energy use conform to a Leontief production function and thus are not substitutable (Fig. 8c). A unit use of fossil energy for production (X_F) is forced to generate a fixed amount of CO_2 (X_E), applying a constant emission factor by fuel type and sector. When an emission cap is imposed, firms will respond to the shock by switching to less carbon-intensive fuels, substituting capital for fossil energy (i.e., adoption of less carbon-intensive technologies), and/or reducing energy consumption. Carbon capture and storage is modeled as a backstop technology, so that it comes into play only when a policy constraint drives up the prices of conventional energy inputs to the level where associated capital investment can be justified.

Three policy scenarios, shown in Table 4, are developed for our analysis, and their simulation results are compared with the BAU case. First, the NO_x and SO_2

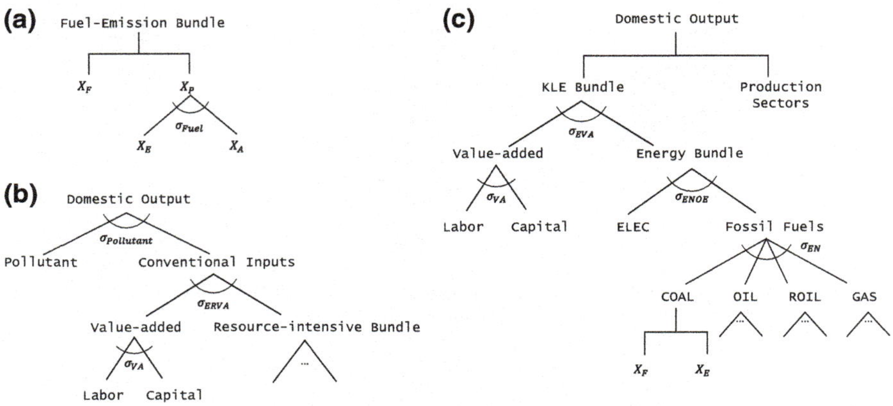

Fig. 8 Pollution abatement and carbon emission structure in EPPA5: (**a**) fuel-related pollutant emissions, (**b**) non-fuel-related pollutant emissions, (**c**) carbon emissions (Source: Modified from Nam et al. (2014), p. 188; reprinted with permission from Elsevier)

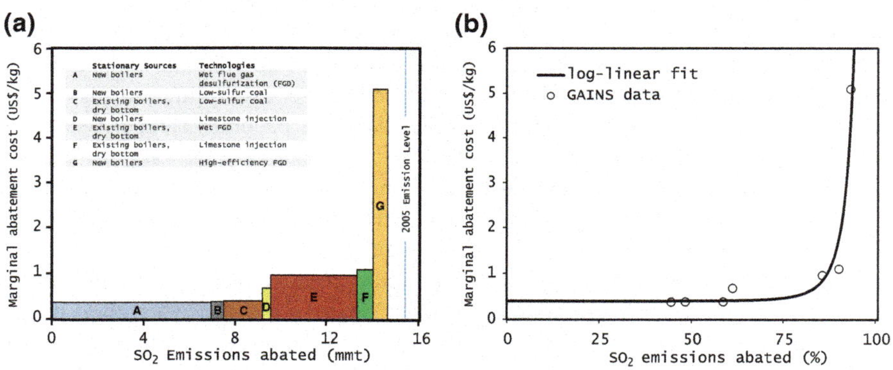

Fig. 9 MAC curve for SO_2 emissions from coal-fired power plants in China, 2005: (**a**) abatement opportunities identified by GAINS, (**b**) estimated MAC curve (Source: Nam et al. (2014), p. 201; reprinted with permission from Elsevier)

Table 4 Policy scenarios

Scenario	Policy goals
NO_x and SO_2 control only	NO_x emissions in 2015 are 10 % below the 2010 level
	SO_2 emissions in 2015 are 8 % below the 2010 level
	NO_x and SO_2 emissions are fixed at the 2015 levels till 2030
CO_2 control only	CO_2 intensity is reduced by 17 % every 5 years
NO_x, SO_2, and CO_2 control	Pollution abatement goals in NO_x and SO_2 control-only scenario and CO_2 intensity reduction goals in CO_2 control-only scenario are enforced at the same time

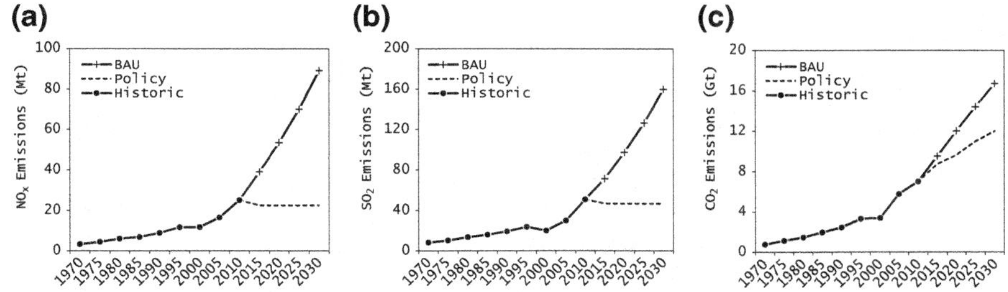

Fig. 10 Historic and future emissions trend in China, 1970–2030: (**a**) NO_x, (**b**) SO_2, (**c**) CO_2 (Source: Created by the author; historic NO_x and SO_2 data from the EDAGAR dataset; historic CO_2 data from the WDI database)

control-only scenario places caps over NO_x and SO_2 emissions without constraining CO_2. In this scenario, the NO_x and SO_2 levels meet their FYP12 targets by 2015 and are then kept constant until 2030. Second, the CO_2 control-only scenario regulates CO_2 but not NO_x or SO_2. With reference to China's FYP12 and Copenhagen targets, we enforce a 17 % carbon intensity reduction every 5 years till 2030. Finally, the NO_x-SO_2 and CO_2 control scenario imposes regulations on the two air pollutants and CO_2 at the same time. Figure 10 displays gas-specific emission trends by scenario.

3.4 Results

Both pollution and carbon regulations require substantial economic costs. Annual policy compliance costs associated with NO_x and SO_2 reduction, when measured in consumption loss, range from $18 billion in 2015 (0.8 % of the BAU consumption level) to $468 billion in 2030 (9.1 % of the BAU consumption level), presenting an increasing trend over time (Fig. 11a).[6] The increasing trend is due to the growing stringency of the pollution control over time (in terms of absolute pollution abatement required against the BAU level), as shown in Fig. 10a, b. The cost of meeting the CO_2 targets also shows an upward trend, increasing from $3 billion in 2015 (0.1 % of the BAU consumption level) to $80 billion in 2030 (1.6 % of the BAU consumption level). The NO_x and SO_2 abatement requires a greater degree of policy compliance costs than the CO_2 mitigation, as the targets for the former are more stringent than those for the latter (compare Fig. 10a, b with c).

One interesting aspect of the result displayed in Fig. 11a is that the carbon co-benefits of NO_x and SO_2 control are exactly the same as the policy compliance costs required to meet CO_2 intensity reduction goals. In 2030, for example, the

[6] All monetary results displayed in this section are measured in constant 2004 US dollars, unless otherwise specified.

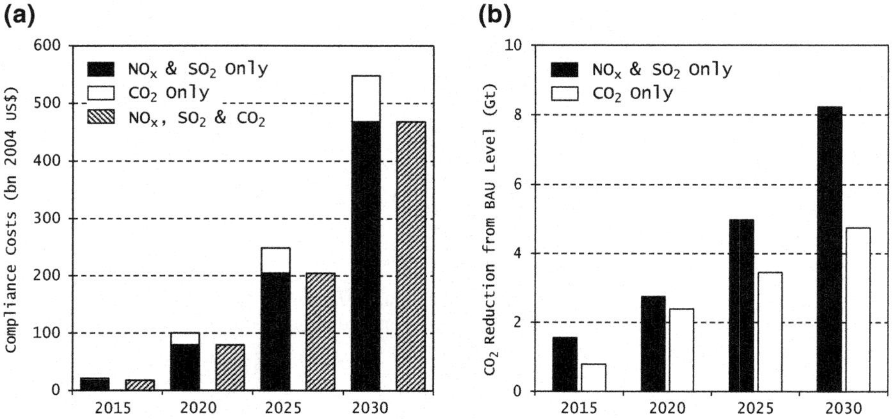

Fig. 11 Compliance costs and CO_2 reductions by scenario, 2015–2030: (**a**) policy compliance costs, (**b**) CO_2 reductions

Chinese economy can save \$80 billion by regulating the two pollutants and CO_2 together (\$468 billion) rather than by regulating them independently without coordination (\$548 billion): herein, the \$80 billion is exactly the same as the policy compliance costs associated with CO_2 mitigation. This tendency—carbon co-benefits identical to the CO_2 regulation costs—is the case for any year between 2015 and 2030. The primary reason is that *unintended* carbon reduction from NO_x and SO_2 abatement exceeds the CO_2 mitigation expected under the proposed carbon intensity reduction targets, making the CO_2 targets not binding. As shown in Fig. 11b, carbon reduction achieved under the NO_x and SO_2 control-only scenario overwhelms the carbon reduction achieved under the CO_2 control-only scenario for all years analyzed. That is, in the presence of the NO_x and SO_2 abatement targets, the CO_2 intensity reduction targets are redundant, as they do not contribute to additional carbon reduction while increasing policy compliance costs.

Out of the 14 production sectors within the model, the power generation (ELEC) and energy-intensive (EINT) industries[7] are affected most by the given emission regulations (Fig. 12). When NO_x and SO_2 emission caps are imposed, for example, 71–78 % of the total NO_x abatement and 69–80 % of the total SO_2 reduction were achieved by the EINT sector alone. The ELEC sector accounted for 9–15 % of the NO_x abatement and 16–24 % of the SO_2 reduction. The sectoral decomposition of carbon reduction shows a slightly different picture, with the relative weights of the two sectors reversed. Under the NO_x and SO_2 constraint, more than half the total unintended CO_2 reduction was from the ELEC sector, while roughly a quarter of it was achieved by the EINT sector.

[7] In EPPA, the EINT sector includes industries engaged in production of paper products, chemical products, ferrous and nonferrous metal products, metal products, and mineral products.

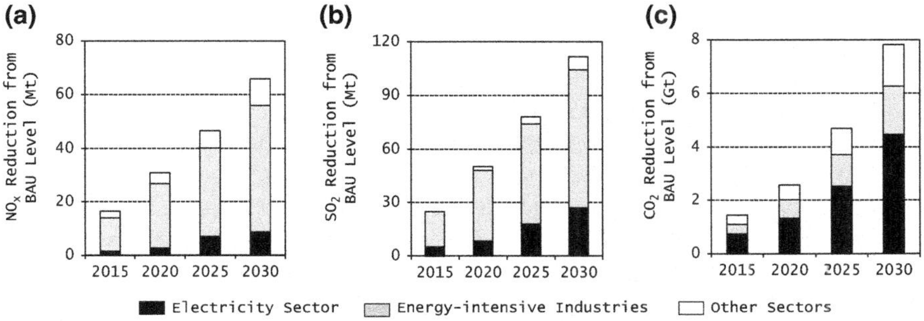

Fig. 12 Decomposition of emission reduction under the NO_x and SO_2 control-only scenario by sector, 2015–2030: (**a**) NO_x reduction, (**b**) SO_2 reduction, (**c**) CO_2 reduction

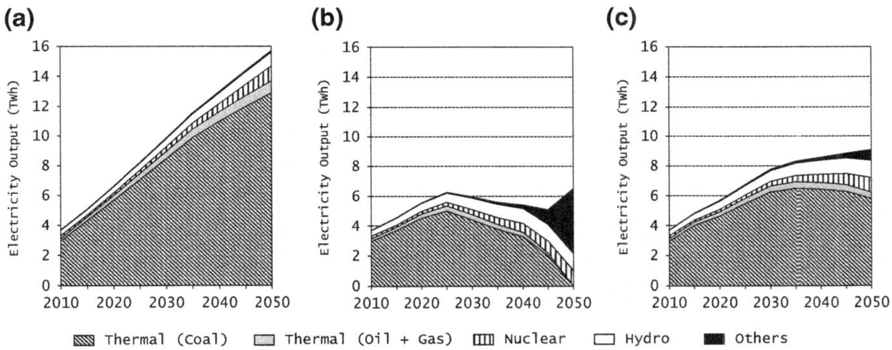

Fig. 13 Electricity output mix by scenario, 2010–2030: (**a**) BAU, (**b**) NO_x and SO_2 control only, (**c**) CO_2 control only

The pollution and carbon regulations also have huge impacts on China's electricity output and its mix. When our simulations are extended to 2050, China's total electricity output during the period will be cut by 10–65 % from the BAU level under the NO_x and SO_2 control-only scenario (Fig. 13a, b). Until 2025, China's ELEC sector can meet the reduction targets while still allowing increased use of coal, by extending the adoption of end-of-pipe control measures. Around at this time point, however, the marginal cost of employing such measures will be excessively high due to exhausted abatement opportunities, requiring less use of coal. Accordingly, coal will be completely phased out of the electricity market by 2050. Cleaner energy sources—wind, in particular—will increasingly replace coal-fired power plants, but their extensive market penetration will not occur until 2050, creating a supply shock between 2040 and 2050. Herein, gradual market penetration of cleaner energy technologies is primarily due to a limited initial pool of the local resources and capabilities required for the immediate and extensive adoption of the technologies at market-competitive prices (Jacoby et al. 2006). The stringency of the pollution abatement targets drives up the turnover rate of capital stock in China's ELEC sector to the level where retirement of conventional fossil energy technologies exceeds the pace at which cleaner technologies come into play.

Under the CO_2 control-only scenario, electricity output cut ranges between 5 and 42 % of the BAU levels (Fig. 13c). China's ELEC sector can meet the CO_2 intensity reduction targets while allowing increased coal use until 2035. After then, coal use will gradually decrease, and a greater role will be played by hydro, nuclear, and renewables. The transition to less carbon-intensive energy sources is much smoother under this scenario than under the pollution abatement scenario, reflecting its relatively modest reduction goals.

4 Summary and Conclusions

Korea regulates air quality and GHGs with explicit policy targets, but rather independently. The nationwide air quality standards cover most of the critical conventional pollutants, and even tougher antipollution regulations, including strict emission caps, are imposed on the Capital Region, requiring a $\geq$44 % reduction from the BAU emission levels by 2024. In addition, Korea has recently launched a nationwide cap-and-trade system to regulate GHG emissions. The near-term goal is to reduce GHG emissions by 30 % or 233 $MtCO_2e$, compared with the BAU level, by 2020. These targets may be achieved at lower costs when the two regulatory schemes are tightly coordinated, given the substantial overlap in emission sources. As of 2011, for example, fossil energy sources fulfilled >80 % of Korea's total energy demand, and their combustion accounted for the majority of air pollutant (e.g., 82 % of SO_x and 95 % of NO_x) and GHG emissions (e.g., 84 % of CO_2). Similarly, three sectors—power stations, industrial combustion, and transportation—accounted for $\geq$60 % of NO_x, SO_x, and CO_2 emissions in the same year. The existing regulations, however, tend to overlook such potential cross effects between pollution and GHG control.

The case of China may give us a rough idea about the magnitude of the potential synergy, both in monetary and physical terms. When focusing on NO_x, SO_2, and CO_2, we find that the potential synergy is substantially large in China. In 2015 alone, for example, carbon co-benefits of meeting the NO_x and SO_2 reduction targets stated in FYP12 are estimated to be $3 billion in saved compliance costs or 1.4 Gt of unintended CO_2 reduction. This 1.4 Gt of CO_2 reduction is equivalent to a 20 % reduction from China's 2010 CO_2 intensity level; in other words, China can overachieve its official 17 % carbon intensity reduction goal by simply meeting the NO_x and SO_2 targets. Accordingly, the CO_2 intensity targets do not bind, and the $3 billion required to meet the goals remains unnecessary policy compliance costs.

Several implications can be drawn for Korea from the Chinese case. First, synergistic interactions between pollution and carbon reduction policies should not be overlooked. As demonstrated in China's case, the regulations aiming at

NO_x and SO_2 reduction initially expand adoption of end-of-pipe pollution abatement technologies, generating limited impacts on carbon emissions. However, the growing exhaustion of remaining abatement opportunities will drive up the marginal abatement cost at a quasi-geometrical rate, and firms will eventually be required to reduce the use of pollution-intensive fuels. This will result in a substantial amount of CO_2 reduction, as pollution-intensive fuels, in general, have a high carbon emission factor as well. From this standpoint, pollution abatement can be viewed as an indirect carbon mitigation strategy.

Second, actualizing the synergy requires a careful coordination between the two policies. As discussed earlier, China is not likely to take the full advantage of its potential synergy since its pollution and carbon abatement regulations are not jointly optimized. In particular, its carbon intensity reduction targets are too modest, compared with its pollution abatement targets, and thus do not bind without net contribution to carbon mitigation. Regulations designed to meet the targets, however, still require substantial compliance costs, resulting in increased market inefficiency without obvious policy benefits. Similar to China, Korea has also introduced explicit emission reduction targets for air pollutants and GHGs at the sectoral and regional levels. Improved coordination between them will help Korea achieve the reduction goals at lower costs. Finally, clear long-term reduction goals and associated policy incentives can reduce the magnitude of a potential supply shock in the energy sector. Control over air pollutant and carbon emissions primarily constrains the electricity sector, as well as a set of energy-intensive industries, and promotes an electricity output mix increasingly biased toward nonfossil energy sources. A likely consequence of this transition, as exemplified by the Chinese case, is a temporary economic shock, caused by a mismatch of the paces at which old capital vintage is retired and advanced energy technologies come online. Avoiding this situation will require clear and consistent long-term goals and associated policy incentives. In their presence, economic agents will be encouraged to act early and possess more forward-looking behavior by placing greater emphasis on investment in advanced energy technologies rather than adoption of incremental control measures.

References

Agee, M.D., S.E. Atkinson, T.D. Crocker, and J.W. Williams. 2014. Non-separable pollution control: Implications for a CO_2 emissions Cap and trade system. *Resource and Energy Economics* 36(1): 64–82.

Bollen, J., B. van der Zwaan, C. Brink, and H. Eerens. 2009. Local air pollution and global climate change: A combined cost-benefit analysis. *Resource and Energy Economics* 31(3): 161–181.

Chae, Y. 2010. Co-benefit analysis of an air quality management plan and greenhouse gas reduction strategies in the Seoul metropolitan area. *Environmental Science and Policy* 13: 205–216.

Chae, Y., and J. Park. 2011. Quantifying costs and benefits of integrated environmental strategies of air quality management and greenhouse gas reduction in the Seoul metropolitan area. *Energy Policy* 39(9): 5296–5308.

Chen, Y.-H.H., S. Paltsev, J.M. Reilly, J.F. Morris, and M. Babiker. 2015. *The MIT EPPA6 model: Economic growth, energy use, and food consumption, MIT JPSPGC report 278*. Cambridge, MA: MIT Joint Program on the Science and Policy of Global Change.

Government of the Republic of Korea. 2014. Roadmap to meeting the national greenhouse gas reduction goals. http://www.greengrowth.go.kr/cmmn/downloadFile.do?file_seq=5&file_sn=2. Accessed 1 Aug 2015.

Jacoby, H.D., J.M. Reilly, J.R. McFarland, and S. Paltsev. 2006. Technology and technical change in the MIT EPPA model. *Energy Economics* 28: 610–631.

Liu, Z., D. Guan, D. Crawford-Brown, Q. Zhang, K. He, and J. Liu. 2013. Energy policy: A low-carbon road map for China. *Nature* 500(7461): 143–145.

Matus, K., K.-M. Nam, N.E. Selin, L.N. Lamsal, J.M. Reilly, and S. Paltsev. 2012. Health damages from air pollution in China. *Global Environmental Change* 22(1): 55–66.

Ministry of Environment (MOE), R.O.K. 2010. *Revised basic plan for Seoul metropolitan air quality management (2005–2014)*. http://www.me.go.kr/mamo/file/readDownloadFile.do;jsessionid=bS36IrXDcMZYmEqOGzIunpipdrKDkyuIcRbEdxLVUTiDrQFjSWhiovM6Uyj7Gp5i.meweb1vhost_servlet_engine1?fileId=84340&fileSeq=1. Accessed 1 Aug 2015.

Ministry of Environment (MOE), R.O.K. 2013. *The second basic plan for Seoul metropolitan air quality management (2015–2024)*. http://www.me.go.kr/mamo/file/readDownloadFile.do?fileId=97490&fileSeq=1. Accessed 1 Aug 2015.

Ministry of Environmental Protection (MEP), P.R.C. 2012. *Ambient air quality standards (GB 3095–2012)*. http://kjs.mep.gov.cn/hjbhbz/bzwb/dqhjbh/dqhjzlbz/201203/t20120302_224165.htm. Accessed 1 Aug 2015.

Morgenstern, R., A. Krupnick, and X. Zhang. 2004. The ancillary carbon benefits of SO_2 reductions from a small-boiler policy in Taiyuan, PRC. *Journal of Environment and Development* 13(2): 140–155.

Nam, K.-M., C.J. Waugh, S. Paltsev, J.M. Reilly, and V.J. Karplus. 2013. Climate co-benefits of tighter SO_2 and NO_x regulations in China. *Global Environmental Change* 23(6): 1648–1661.

Nam, K.-M., C.J. Waugh, S. Paltsev, J.M. Reilly, and V.J. Karplus. 2014. Synergy between pollution control and climate policy: Comparing China and the United States. *Energy Economics* 46: 186–201.

Nielsen, C.P., and M.S. Ho. 2013. Atmospheric environment in China: Introduction and research review. In *Clearing skies over China: Reconciling air quality, climate, and economic goals*, ed. C.P. Nielsen and M.S. Ho. Cambridge, MA: MIT Press.

Paltsev, S., J.M. Reilly, H.D. Jacoby, R.S. Eckaus, J. McFarland, M. Sarofim, M. Asadoorian, and M. Babiker. 2005. *The MIT emissions prediction and policy analysis (EPPA) model: Version 4, MIT JPSPGC report 125*. Cambridge, MA: MIT Joint Program on the Science and Policy of Global Change.

Shindell, D., G. Faluvegi, M. Walsh, S.C. Anenberg, R. van Dingenen, N.Z. Muller, J. Austin, D. Koch, and G. Milly. 2011. Climate, health, agricultural and economic impacts of tighter vehicle-emission standards. *Nature Climate Change* 1: 59–66.

Wang, S., and J. Hao. 2012. Air quality management in china: Issues, challenges, and options. *Journal of Environmental Sciences* 24(1): 2–13.

World Bank, and SEPA (P.R.C. State Environmental Protection Administration). 2007. *Cost of pollution in China: Economic estimates of physical damages*. Washington, DC: World Bank.

World Health Organization (WHO). 2006. *WHO air quality guidelines for particulate matter, ozone, nitrogen dioxide and sulfur dioxide: global update 2005*. Geneva: WHO Press.

Xu, Y., and T. Masui. 2009. Local air pollutant emission reduction and ancillary carbon benefits of SO_2 control policies: Application of AIM/CGE model to China. *European Journal of Operational Research* 198: 315–325.

Effects of the Project Investments and Valuation of the Water Quality Improvement of the River Taehwa in Ulsan, Korea

Jae Hong Kim

Abstract This study analyzes the effects of the project investments on the river water quality improvement and also provides contingent valuation estimates of household's willingness to pay (WTP) to continue public investment to the river water quality improvement and maintenance. The estimation results using the OLS regression models with correction of autocorrelation show that the household soil pipe connection project with investment of 26.7 billion KRW has reduced 1.68 ppm in BOD and that the project dredging sediments at the river bottom with investment of 16 billion KRW has resulted in the decrease of 1.12 ppm in BOD at the downstream of the River Taehwa. Using a contingent valuation method with multiple choices in consideration of respondent's uncertainty, the estimation results of four logit models show that truncated mean household's WTP is estimated from 1224.7 KRW to 2747 KRW for the respective four models. The present values of total social benefits in the Ulsan Metropolitan City are estimated from 196 billion KRW to 441 billion KRW for the respective four models, when applying the 3 % discount rate.

Keywords Water quality improvement • Soil pipe connection • Dredging sediment • Contingent valuation • Preference uncertainty • Correction of autocorrelation

1 Introduction

The Ulsan Metropolitan City is the largest industrial city in Korea. The city had been stigmatized as a pollution city with heavy industrialization since 1970s. Recently, however, the city has made a model case in Korea in terms of the transformation from a pollution city to an eco-city. The water quality improvement

J.H. Kim (✉)
Department of Public Administration, School of Social Sciences, University of Ulsan, Ulsan, Korea
e-mail: jhkim@ulsan.ac.kr

E. Kim, B.H.S. Kim (eds.), *Quantitative Regional Economic and Environmental Analysis for Sustainability in Korea*, New Frontiers in Regional Science: Asian Perspectives 25, DOI 10.1007/978-981-10-0300-4_12

of the River Taehwa, which is running midst of the city, is one of the dramatic environmental achievements. The river water quality at the downstream area had been improved from 11.7 ppm in 1991 to 1.7 ppm in 2007 in terms of BOD. In order to improve the river water quality, the city had already invested about 130 billion KRW (around US$ 93 million) during the period of 1998–2007 and has a plan to invest more than 133 billion KRW (around US$ 10.6 million) after 2008. For all such a large amount of public investment to the river water quality improvement projects, up to date, there was neither attempt except one (Kim 2009) to analyze the effects of the project investments on the river water quality improvement, nor to evaluate the social benefits from the improvement of the water quality in the River Taehwa except two studies (Kim 2007, 2009).

This study analyzes not only the effects of the project investments on the river water quality improvement but also provides contingent valuation estimates of household's willingness to pay (WTP) to continue public investment to the river water quality improvement and maintenance. However, the main focus is placed on the estimation of the WTP and social benefits from the river water quality improvement and maintenance. This study utilizes the internal data from the Department of Environment Policy of the Ulsan Metropolitan City with monthly measured records of river water quality at the four measurement points for 10 years from 1998 to 2007 and uses a TSCS (time series cross section) model with correction of autocorrelation to analyze the effects of the project investments on the river water quality improvement by the four measurement points. In order to estimate citizen's WTP and the social benefits from improvement of river water quality, this study uses a contingent valuation method with preference uncertainty on the basis of a multiple choice WTP survey with 397 respondents of the Ulsan Metropolitan City.

The rest of this paper is organized as follows: Section 2 introduces the trend of river water quality and the project investment for water quality improvement at the downstream of the River Taehwa and analyzes the effects of the project investment on water quality improvement by the measurement points. Section 3 outlines the valuing methods for nonmarket goods and proposes a contingent valuation method with multiple choices in consideration of preference uncertainty utilized in this study. Section 4 describes the survey design and the collected data for a contingent valuation study. Section 5 analyzes the estimation results and compares the WTP estimates of the four types of logit models. Section 6 summarizes the major findings and highlights some policy implications.

2 Water Quality Improvement Project of the River Taehwa

2.1 Trend of Water Quality of the River Taehwa

As shown in Table 1, the river water quality measured in biological oxygen demand (BOD) at the downstream of the River Taehwa had been improved from 6.47 ppm in 1998 to 1.66 ppm in 2007 in terms of the arithmetic mean. The variation of the water quality also had been gradually decreased. The four water quality measurement points at the downstream of the River Taehwa are located in lines from Samho at the uppermost through Taehwa and Haksung to Myungchon at the river mouth. The river water quality at the Samho measurement point had been gradually improved from 3.21 ppm in 2000 to 0.98 ppm in 2007. However, at the Taehwa measurement point, which is the starting point of the population concentrated urban area, the water quality had not been significantly improved from 1998 to 2006, but conspicuously improved into 2.00 ppm in 2007 when both the household soil pipe connection project and the project dredging sediment of the river bottom were accomplished. At the Haksung measurement point, the water quality had been dramatically improved from 15.34 ppm in 1998 to 1.88 ppm in 2007, and the variation has also been going down since 1998. The water quality at the downmost Myungchon measurement point had been also gradually improved from 4.89 ppm in 1999 to 1.77 ppm in 2007. This improvement of the water quality at the downstream of the River Taehwa has resulted from the various water quality improvement projects continuously carried into execution by the Ulsan Metropolitan City since 1995. The city had already invested about 130 billion KRW (around US$ 93 million) during the period of 1998–2007 for the improvement of water quality at the downstream of the River Taehwa. Especially, the household soil pipe connection project with investment of 45 billion KRW (26.6 billion KRW during 1998–2007) and the dredging sediment project with investment of 47 billion KRW (16.0 billion KRW in the part of the downstream during 2004–2007) are noteworthy in that the water quality at all the measure points of the downstream dramatically improved in 2007 when the two projects were finished by September and July, respectively. Thus in the next section, the effects of the two projects on the water quality improvement at each measurement points at the downstream of the River Taehwa will be analyzed.

2.2 Analysis of the Effects of the Project Investments on River Water Quality Improvement

2.2.1 Empirical Model

This study uses a regression analysis using BOD at each measurement point as a dependent variable and the amount of the project investments at each point, the BOD at the next upper measure point, and seasonal dummies as dependent variables in order to estimate the effects of independent variables on the dependent variable. In a regression equation $\boldsymbol{Q_{it}} = \boldsymbol{\alpha} + \boldsymbol{\beta_1 X_1} + \boldsymbol{\beta_2 X_2} + \boldsymbol{\beta_3 Q_{(1+i)t}} + \boldsymbol{\beta_k D_k} + \boldsymbol{\varepsilon}$, let $\boldsymbol{Q_{it}}$ be BOD at the measure point i and time t; $\boldsymbol{X_1}$ and $\boldsymbol{X_2}$ be the cumulative amount of the project investments until time t for the household soil pipe connection and dredging sediment, respectively; $\boldsymbol{Q_{(1+i)t}}$ be BOD at the next upper measure point i and time t; and $\boldsymbol{D_k}$ be the seasonal dummies, and then $\boldsymbol{\beta}$'s explain the effects on the

Table 1 Trend of river water quality (BOD) at the measurement points of the downstream in the River Taehwa from 1998 to 2007 (unit: mg/l)

Measurement points		1998	1999	2000	2001	2002	2003	2004	2005	2006	2007
Samho	Mean	1.86	2.37	3.21	2.38	1.62	1.05	1.19	1.08	1.32	0.98
	Std	1.00	1.48	2.02	1.31	0.93	0.34	0.72	0.47	0.68	0.51
	Min/max	0.7/3.8	0.7/4.6	0.8/6.3	0.8/4.5	0.6/3.4	0.1/1.4	0.2/3.0	0.2/2.0	0.3/2.5	0.1/1.8
Taehwa	Mean	4.21	4.10	5.00	6.47	4.88	3.38	5.08	4.03	6.01	2.00
	Std	1.57	2.41	2.72	4.57	3.41	3.37	4.64	2.66	8.68	1.03
	Min/max	1.8/6.8	1.3/8.7	1.5/9.5	2.0/18.5	1.7/11.9	0.8/11.4	0.9/13.2	0.7/8.3	1.1/31.8	0.8/4.0
Haksung	Mean	15.34	4.59	7.32	8.18	5.60	3.34	3.96	2.98	3.22	1.88
	Std	17.51	3.33	7.20	6.46	4.84	2.16	2.03	1.30	1.96	0.71
	Min/max	2.0/61.0	0.9/10.6	1.0/24.0	3.2/22.8	1.6/19.5	0.6/7.0	1.7/9.0	1.7/5.6	1.4/8.0	0.2/2.9
Myungchon	Mean	4.84	4.89	4.23	4.58	5.43	2.81	2.72	2.55	2.43	1.77
	Std	1.66	2.85	2.20	2.33	3.16	1.22	1.38	1.11	1.16	1.12
	Min/max	2.7/6.9	1.6/9.5	1.5/8.4	2.4/10.8	2.0/12.3	1.1/5.5	1.2/5.2	1.2/4.5	1.1/5.4	0.8/4.4
Downstream Total	Mean	6.47	3.99	4.94	5.40	4.38	2.65	3.24	2.66	3.22	1.66
	Std	9.80	2.71	4.27	4.59	3.66	2.25	2.95	1.88	4.69	0.94
	Min/max	0.7/61.0	0.7/10.6	0.8/24.0	0.8/22.8	0.6/19.5	0.1/11.4	0.2/13.2	0.2/8.3	0.3/31.8	0.1/4.4

Sources: Author's calculation from the internal data of monthly water quality measures at the each measurement points at the downstream of the River Taehwa during 1998–2007 provided from the Ulsan Metropolitan City (2008)

water quality improvement of one unit increase of each independent variable. For an empirical model in this study, a semilog form transposes the dependent variable with a natural log as follows:

$$\mathbf{lnBOD} = f(\mathbf{Soil\ Pipe,\ Sediment,\ Upper\ BOD,\ Seasonal\ Dummies}) \tag{1}$$

When estimating Eq. (1) with OLS, the estimated coefficients may be biased because of autocorrelation of error terms due to the characteristics of the time-series data. Therefore, in the case of existence of autocorrelation of error terms, Eq. (1) should be corrected by using the coefficient of autocorrelation ρ calculated from the Durbin-Watson statistics, $d = 2(1-\rho)$. Thus, first-order autocorrelation regression (AR1) using the maximum likelihood estimation instead of OLS is applied in this study. The estimation procedure of AR1 is as follows: Let lnBOD in Eq. (1) be Y_t, and then the estimation model is given as Eq. (2).

$$\begin{aligned} Y_t &= \beta' X_t + \varepsilon_t \\ \varepsilon_t &= \rho \varepsilon_{t-1} + \mu_t \end{aligned} \tag{2}$$

In the next step, the autocorrelation coefficient is calculated from the Durbin-Watson statistics from the OLS estimation of Eq. (2). Then Eq. (2) is transposed into Eq. (3) for AR1 estimation.

$$\begin{aligned} & Y_1^* = (1-\rho^2)^{1/2} Y_1, \quad X_1^* = (1-\rho^2)^{1/2} X_1 \\ & Y_t^* = Y_t - \rho Y_{t-1} = \beta^* (X_t - \rho X_{t-1}) + \upsilon_t \end{aligned} \tag{3}$$

In this study, Eq. (3) is estimated by feasible generalized least square (FGLS) using the maximum likelihood estimation. Equation (3) takes advantage of avoiding loss of the first observation, which occurs in the two-step OLS using the Cochrane-Orcutt method.

Even though Eq. (3) is appropriate for estimation of each measurement point, it has some problems in estimation for the whole river model including all of the four measurement points. Those problems include spatial heteroskedasticity and correlation between measurement points besides autocorrelation of error terms already mentioned above. In order to solve the estimation problems, this study adopts a TSCS (time series cross section) model, which is suggested by Greene (2003). Thus our final model is given in Eq. (4):

$$\begin{aligned} & Y_{1i}^* = (1-\rho^2)^{1/2} Y_{1i}, \quad X_{1i}^* = (1-\rho^2)^{1/2} X_{1i} \\ & Y_{ti}^* = Y_{ti} - \rho Y_{t-1i} = \beta^* (X_{ti} - \rho X_{t-1i}) + \upsilon_{ti} \end{aligned} \tag{4}$$

where t indicates the time index, i means the space index, and β^* is estimated identically regardless of time and space. Equation (4) is estimated by the three-step FGLS using the maximum likelihood method in order to consider spatial heteroskedasticity, spatial correlation, and autocorrelation of error terms at the same time.

2.3 Estimation Results and Analysis

When estimating Eq. (1) with OLS, autocorrelation of error terms was found in the downstream model but that is rejected in the other four models. Thus the autocorrelation corrected model was applied for the downstream model. Table 2 summarizes the estimation results for five models.

In the downstream model, upper BOD and soil pipe are statistically significant at the 5 % level, and sediment is statistically significant at the 10 % level. These results can be explained as follows: The 1 ppm decrease of upper BOD resulted in the 2.8 % (0.098 ppm) decrease of the downstream BOD at the mean. The one billion of investment on the household soil pipe connection project diminished the 1.8 % (0.063 ppm) of the downstream BOD. This means that the 1.68 ppm decrease of the downstream BOD resulted from the total investment of 26.6 billion KRW on the household soil pipe connection project during the period of January1998–September 2007. The one billion of investment on the project dredging sediment lessened the 2.0 % (0.07 ppm) of the downstream BOD. This represents that the 1.12 ppm decrease of the downstream BOD was caused by the total investment of 16.0 billion KRW on the project dredging sediment during the period of April 2004–February 2007.

The estimation results from the four measurement points are as follows: Upper BOD is statistically significant at least at 5 % level in the three measurement points except the Haksung point. Soil pipe is statistically significant at least at 10 % level in the three measurement points except the Taehwa point. Sediment is statistically significant at 5 % level in both Taehwa and Myungchon. The insignificant soil pipe and sediment may arise from the multi-collinearity problem due to the overlapping period of investment for the two projects.

The effect of the household soil pipe connection project seems to be the greatest in the Haksung point showing the 1.98 ppm decrease of BOD at the point during the same period. The project dredging sediment made the greatest effect on the decrease of BOD at the point by 2.20 ppm.

As also shown in Table 2, the river water quality at the downstream of the River Taehwa is seasonally different with statistical significance. That is the best in October and the worst in February in terms of monthly means for 10 years from 1998 to 2007. The estimates of the monthly dummies show that the water quality during the water shortage period from December to June is significantly worse than that of October, but that of the period from July to November is not statistically different from that of October at the overall downstream. Although the monthly effects show the same pattern for all the four measurement points, the statistical significances are different among the four points. This result indicates that the monthly effects are getting weaker as the measure points are closer to the river mouth because of relatively smaller variation of the volume of water at the lower reaches.

The Ulsan Metropolitan City has a plan to invest more than 133 billion KRW (around US$ 10.6 million) to improve and maintain the steady water quality from 2008. The willingness to pay and social benefits from improvement and maintenance of the steady water quality are estimated in the following sections.

Table 2 Estimation results of the five regression models

	Downstream	Samho	Taehwa	Haksung	Myungchon
Variable	TSCS model	AR1-ML model	AR1-ML model	AR1-ML model	AR1-ML model
Constant	1.133 (5.026)**	0.407 (1.203)	0.590 (1.652)	1.535 (4.334)**	1.520 (5.664)**
Upper BOD	0.028 (3.977)**	0.273 (3.021)**	0.204 (3.538)**	0.024 (1.361)	0.020 (2.772)**
Soil pipe	−0.018 (−2.322)**	−0.030 (−3.407)**	0.005 (0.577)	−0.017 (−1.701)*	−0.017 (−2.377)**
Sediment	−0.020 (−1.810)*	0.006 (0.418)	−0.032 (−2.206)**	−0.025 (−1.545)	−0.023 (−2.005)**
January	0.589 (3.817)**	0.841 (3.107)**	0.631 (2.254)**	0.377 (1.253)	0.359 (1.666)*
February	0.626 (4.041)**	0.897 (3.402)**	0.654 (2.393)**	0.542 (1.666)*	0.357 (1.636)
March	0.535 (3.481)**	0.642 (2.432)**	0.439 (1.607)	0.724 (2.408)**	0.213 (0.975)
April	0.429 (2.763)**	0.583 (2.206)**	0.465 (1.719)*	0.378 (1.196)	0.181 (0.829)
May	0.458 (2.969)**	1.012 (3.812)**	0.346 (1.230)	0.080 (0.261)	0.151 (0.696)
June	0.417 (2.733)**	0.752 (2.793)**	0.081 (0.286)	0.224 (0.751)	0.264 (1.225)
July	0.141 (0.925)	0.734 (2.765)**	−0.385 (−1.408)	0.111 (0.370)	−0.092 (−0.416)
August	0.210 (1.446)	0.548 (2.057)**	−0.127 (−0.472)	0.066 (0.227)	0.214 (0.995)
September	0.029 (0.229)	0.186 (0.696)	−0.326 (−1.224)	−0.041 (−0.142)	0.133 (0.619)
November	0.142 (1.115)	0.226 (0.858)	0.232 (0.873)	0.079 (0.273)	0.012 (0.056)
December	0.273 (1.878)*	0.280 (1.061)	0.339 (1.274)	0.396 (1.359)	−0.017 (−0.081)
Pseudo R^2	0.183	0.447	0.439	0.321	0.438
AR-ML DW	1.987	1.916	2.059	2.079	2.027
Sample size	476	119	119	118	120

Note: Values in parentheses indicate t value; and * and ** represent statistical significances at 10 % and 5 %, respectively

3 Valuation of River Water Quality Improvement

3.1 Contingent Valuation Method (CVM) Using a Utility Difference Model

CVM has been a popular technique for eliciting the value of nonmarket goods, using a Hicksian welfare function derived from a hypothetical market. Recently, a utility difference model proposed by Hanemann (1984) and an expenditure function model suggested by Cameron (1988) are widely used to estimate welfare functions and calculate welfare measure. McConell (1990) shows both two models are dual to each other, and the choice between two models is a matter of style as much as of known defects and merits. This study uses a CVM based on a utility difference model.

In the referendum type of CVM, a respondent accepts (=1) or rejects (=0) to a given bidding price (A). This leads to a kind of binary choice model. Let $V(\cdot)$ denote respondent's indirect utility function, Y and S indicate income and individual characteristics. A respondent will accept the bidding price (A) if acceptance (=1) provides the respondent with more utility than rejection (=0) does. This can be expressed as follows:

This can be expressed as follows:

$$V(1, Y - A, S) + \varepsilon_1 > V(0, Y, S) + \varepsilon_0 \tag{5}$$

where ε_1 and ε_0 are assumed as random errors with zero mean. Then let the probability to accept a given bidding price (A) be π_1 and the probability to reject be π_0:

$$\begin{aligned} \pi_1 &= P[V(1, Y - A, S) - V(0, Y, S) > \varepsilon_0 - \varepsilon_1] \\ \pi_0 &= 1 - \pi_1 \end{aligned} \tag{6}$$

where $P(\cdot)$ denotes a probability function. Then Eq. (6) be expressed as a cumulative density function:

$$\pi_1 = P(\Delta V > \theta) = F_\theta(\Delta V) \tag{7}$$

where $\Delta V = V(1, Y - A, S) - V(0, Y, S)$ and $F_\theta(\cdot)$ denotes a cumulative density function of. Now if we assume $F_\theta(\cdot)$ is a logistic probability function, a log-likelihood function is derived to estimate parameters of ΔV using a maximum likelihood estimation as follows:

$$\ln L = \sum_1^n \left[I_i^Y \ln F_\theta(\Delta V) + I_i^N \ln\left(1 - F_\theta(\Delta V)\right)\right] \tag{8}$$

where $I_i^Y = 1$ if a bidding price is accepted, 0 otherwise, and $I_i^N = 1$ if rejected, 0 otherwise. If we assume ΔV is a linear function including only a random variable, bidding price (A), then $\Delta V = a - bA$, where $b > 0$. Following Hanemann (1984), the expected value (mean) and median of willingness to pay (*WTP*) are evaluated as follows:

$$E(WTP) = WTP_{\text{mean}} = WTP_{\text{median}} = \frac{a}{b} \tag{9}$$

While Eq. (9) considers respondent's negative *WTP*, if assumed *WTP* is greater than or equal to zero, that is, *WTP* is truncated at zero, the truncated mean is given as follows:

$$WTP_{\text{truncated}} = \frac{1}{b}\ln(1 + \exp(a)) \tag{10}$$

The confidence interval of *WTP* can be obtained using a parametric bootstrap method introduced by Krinsky and Robb (1986, 1990) and a delta method based on the Monte Carlo simulation.

3.2 Contingent Valuation Method with Preference Uncertainty

While there are many different contingent valuation elicitation formats such as open-ended, bidding games, payment card, contingent ranking, dichotomous choice, and so on, the guidelines proposed by National Oceanic and Atmospheric Administration (NOAA) panel recommended that "the valuation question should be posed as a vote on a referendum" and "a 'no-answer' option should be explicitly allowed in addition to the 'yes' and 'no' vote options on the main valuation question" (Arrow et al. 1993). However, implementing the "no-answer" option in a contingent valuation survey has a potentially serious cost because of the loss of choice information from a portion of the sample (Wang 1997).

Thus a practical issue arises to treat "no-answer" responses as a form of "not sure" or "don't know." One practice is to drop the "no-answer" responses from the data set, or to treat "no-answer" responses simply as "no" as shown in Carson et al. (1994). Another strategy is to treat "not sure" or "don't know" answers as a category between "yes" and "no" reflecting respondent's choice uncertainties involved in valuation, as shown in Wang (1997). Furthermore, Ready et al. (1995) and Chang et al. (2005, 2007) introduced a polychotomous choice (PC) model, and Welsh and Poe (1998) introduced a multiple bounded discrete choice (MBDC) model, which allows respondents to express their level of voting [un]certainty for a wide range of referendum thresholds. In MBDC model, at each referendum threshold, the respondent is asked, using a scale from "definitely no" to "definitely yes," to indicate how he/she would vote if passage of the referendum costs them that amount (Welsh and Poe 1998). These kinds of CV elicitation formats are derived from endeavors to incorporate individual preference uncertainty, even though those uncertain response options cannot resolve perfectly the problem of the "no-answer" or "don't know" category in terms of conceptualization and methodology Alberini et al. (2003).

Another stream dealing with preference uncertainty is to use a follow-up [un] certainty question to dichotomous choice (DC) model without offering "no-answer" option. Li and Mattsson (1995) elicit a post-decisional confidence measure (0–100 %) by a follow-up debriefing question in addition to a standard dichotomous choice contingent valuation model in order to incorporate respondents' preference [un]certainty. Whitehead et al. (1998), Loomis and Ekstrand (1998), and Berrens et al. (2002) also introduce various alternative CV elicitation formats for incorporating respondent uncertainty and show more improved estimation results from the models with respondent uncertainty than a type of "one-shot" DC models.

3.3 Proposed Models with Preference Uncertainty

In this study, we utilize two types of CV models for incorporating respondent's preference uncertainty such as a PC model and a probability-driven weighted model. In our PC model, we use a polychotomous choice (PC) format with four categories of "definitely yes" (DFYES), "maybe yes" (MBYES), "maybe no" (MBNO), and "definitely no" (DFNO). Analysis of WTP is conducted using the maximum likelihood interval modeling approach. In the analysis of WTP, two empirical models (DFYES model and MBYES model) are provided in this study: In the DFYES model, "definitely yes" is coded as 1 and otherwise 0, and in the MBYES model, "definitely yes" and "maybe yes" are coded 1 and otherwise 0. Two models are estimated as a standard logistic model. Comparison between two estimated WTP will provide information of the changes of WTP and efficiency of estimates over strength of respondent [un]certainty.

In our probability-driven weighted (PW) model, we do not use a DC format with a follow-up certainty question. Instead we take an arbitrary weight for each PC category in our PC format as follows: DFYES = 1, MBYES = 0.75, MBNO = 0.75, and DFNO = 1. This model takes some advantages as follows: First, this method reduces time and cost involved in a new sample survey for a DC with follow-up certainty question. Second, the defaults of a weight-free nature of the PC model may be reduced through the weight for each category reflecting strength of certainty in the PC format, even though the weight for each category is arbitrarily taken. Third, the arbitrary weighting for certainty in a PC format could incorporate respondent uncertainty better than a DC with the follow-up question, because the degree of certainty in a DC with the follow-up question tends to be still uncertain.

Analysis of *WTP* in our *PW* model is conducted analogously to the method of Loomis and Ekstrand (1998) and Chang et al. (2005, 2007). Let $F(\cdot)$ denote a logistic probability function, W^{DY}, W^{MY}, W^{MN}, and W^{DN} indicate a weighting index for each category of DFYES, MBYES, MBNO, and DFNO, respectively, and $I^{DY} = 1$ if DFYES, 0 otherwise; $I^{MY} = 1$ if MBYES, 0 otherwise; $I^{MN} = 1$ if MBNO, 0 otherwise; and $I^{DN} = 1$ if DFNO, 0 otherwise. Then the corresponding log-likelihood function is as follows:

$$\ln L = \sum_{1}^{n} \left[\left(W_i^{DY} I_i^{DY} + W_i^{MY} I_i^{MY} \right) \ln F(\cdot) + \left(W_i^{MN} I_i^{MN} + W_i^{DN} I_i^{DN} \right) \ln(1 - F(\cdot)) \right] \tag{11}$$

In this study, two PW models are established such as a symmetric (SYM) model and an asymmetric (ASYM) model as shown in Loomis and Ekstrand (1998). A symmetric model includes the weights for all categories in the PC format; and an asymmetric model excludes the weights for MBNO and DFNO. The log-likelihood function for an asymmetric model is the same as Eq. (11) except $W^{MN} = W^{DN} = 1$.

4 Valuation Experiment

4.1 *Survey Design*

In this study, the questionnaire for the contingent valuation survey includes respondent's individual characteristics such as age, gender, level of education, household's monthly income, residential location, degree of acknowledgement on the River Taehwa Master Plan, degree of satisfaction on water quality of the River Taehwa, and contingent valuation (CV) question. The CV question is designed as a polychotomous choice (PV) format with four categories of "definitely yes" (DFYES), "maybe yes" (MBYES), "maybe no" (MBNO), and "definitely no" (DFNO). In the CV question, the present and past situations of water quality in the downstream of the River Taehwa are described, and then willingness to pay for water quality improvement charges is asked with a PC format. The PC question consists of eight bidding price thresholds, and a respondent receives one of 8 monthly amount ranging from KRW 500 to KRW 5000 (KRW 500, 1000, 1500, 2000, 2500, 3000, 4000, 5000). Respondents' answer to these questions would be used to develop estimates of respondent households' willingness to pay for water quality improvements and preservation in the downstream of the River Taehwa.

4.2 Data Collection and Description

For the data collection, a contingent valuation survey was conducted in order to estimate the economic value of water quality improvements and preservation at the downstream of the River Taehwa. The survey was undertaken by face-to-face visiting interviews for the randomly selected 430 households in the Ulsan Metropolitan City during the third week of November 2008. The survey proceeded as follows: First, a preliminary survey was conducted for 30 households in order to investigate the relevancy of the survey questionnaire. Second, the 430 sample households were proportionally distributed according to the population of the administrative localities in the Ulsan Metropolitan City. Third, one respondent aged over 19 in each household was face-to-face interviewed with a surveyor pre-trained on CVM. All the 430 questionnaires were returned, but the usable responses were 397 due to some missing variables. Table 3 explains selected variables and describes mean values and standard deviations.

It is noteworthy that an epochal increase has been made in the degree of satisfaction with water quality in the downstream of the River Taehwa for the last 10 years. As shown in Table 4, while respondents' answers of "good" and "excellent" to the water quality was 12.7 % in the 1999 survey, those answers present 34.8 % in this survey. This reflects the real improvements in water quality of the River Taehwa during the period.

Table 5 summarizes respondents' answers to the PC questions at different bidding prices. The percentage of positive answers such as DFYES and MBYES decreases as the bidding price increases. This suggests that the bidding price tends to have a significant impact on the positive answers.

Table 3 Description and values of selected variables

Variable	Description	Mean	Std
Gender	Male = 1, female = 0	0.428	0.495
Age	Respondent's age	39.106	11.111
Education	Levels of school completion (middle school or less = 1, high school = 2, college with 2 years = 3, university with 4 years = 4, postgraduate = 5)	2.987	0.988
Income	Household's monthly income (million KRW)	3.454	1.893
Master plan	Degree of acknowledgement on the River Taehwa Master Plan (don't know at all = 1, ···, know very well = 4)	2.068	0.709
Satisfaction	Degree of satisfaction with water quality in the downstream of the River Taehwa (very poor = 1, ···, excellent = 5)	3.146	0.794
Bid	Bidding price for monthly payment (thousand KRW)	2.471	1.419

Table 4 Degree of satisfaction with water quality of the River Taehwa (unit: person, %)

Year	Very poor	Poor	Fair	Good	Excellent	Total
1999	256 (25.6)	365 (36.5)	252 (25.2)	114 (11.4)	12 (1.3)	1000 (100.0)
2008	8 (2.0)	67 (16.9)	190 (47.9)	123 (31.0)	9 (2.3)	397 (100.0)

Table 5 Percentage of respondents vs. monthly surcharge

Bidding price (KRW)	Sample size	Polychotomous choice			
		Definitely yes	Maybe yes	Maybe no	Definitely no
500	52 (100.0 %)	30 (57.7 %)	11 (21.2 %)	8 (15.4 %)	3 (5.8 %)
1000	40 (100.0 %)	18 (45.0 %)	8 (15.0 %)	6 (15.0 %)	10 (25.0 %)
1500	50 (100.0 %)	18 (36.0 %)	8 (16.0 %)	13 (26.0 %)	11 (22.0 %)
2000	49 (100.0 %)	14 (28.6 %)	9 (18.4 %)	11 (22.4 %)	15 (30.6 %)
2500	54 (100.0 %)	14 (25.9 %)	10 (18.5 %)	15 (27.8 %)	15 (27.8 %)
3000	50 (100.0 %)	10 (20.0 %)	9 (18.0 %)	17 (34.0 %)	14 (28.0 %)
4000	53 (100.0 %)	3 (5.7 %)	7 (13.2 %)	26 (49.1 %)	17 (32.1 %)
5000	49 (100.0 %)	2 (4.1 %)	6 (12.2 %)	20 (40.8 %)	21 (42.9 %)
Total	397 (100.0 %)	109 (27.5 %)	66 (16.6 %)	116 (29.2 %)	106 (26.7 %)

5 Analysis of Estimation Results

5.1 *Estimation Results of the CV Models*

As shown in Table 6, the logit estimation results show that bid price and household's income are statistically most significant variables in all the four models. In terms of sociodemographic variables, gender is statistically significant in the only MBYES model among four models. This means male is more willing to pay for the improvement of river water quality than female at least in the MBYES model. The degree of satisfaction on the river water quality (Satisfaction) has the positive relation with the WTP but is statistically significant at the 10 % in the MBYES model only. The WTP tends to increase significantly with age and with higher acknowledgement of the River Taehwa Master Plan (Master_Plan) in the three models except the SYM model. Education is positively related with the WTP but statistically insignificant in the all the models.

5.2 *WTP Estimation Results*

As shown in Table 7, the estimated truncated mean WTP on the basis of the logit estimation results comes to 1223.7 KRW in the DFYES model, 2333.4 KRW in the MBYES model, 2062.1 KRW in the ASYM model, and 2747.5 KRW in the SYM model, respectively. The WTP estimates in the three models except in the DFYES model have no statistically significant differences. The significantly lower WTP in the DFYES model results from the strict exclusion of preference uncertainty in the DFYES model.

Table 6 Logit estimation results

Variable	PC model DFYES	 MBYES	PW model ASYM	 SYM
Constant	−5.330 (−4.325)**	−4.038 (−4.001)**	−4.127 (−4.025)**	−2.421 (−2.666)**
Gender	−0.223 (−0.732)	0.577 (2.250)**	0.402 (1.558)	0.330 (1.405)
Age	0.031 (1.914)*	0.027 (2.040)**	0.026 (1.942)*	0.119 (1.000)
Education	0.225 (1.281)	0.114 (0.791)	0.134 (0.914)	0.038 (0.283)
Master plan	0.414 (1.854)*	0.389 (2.131)**	0.374 (2.011)**	0.275 (1.640)
Satisfaction	0.148 (0.796)	0.284 (1.784)*	0.247 (1.534)	0.214 (1.457)
Income	0.891 (7.536)**	0.672 (7.222)**	0.682 (7.242)**	0.496 (6.159)**
Bid	−0.988 (−7.136)**	−0.768 (−7.346)**	−0.770 (−7.220)**	−0.548 (−6.166)**
Log likelihood	−146.533	−196.878	−192.461	−224.853
Chi squared	173.601	151.026	197.206	112.912
McFadden's R^2	0.372	0.277	0.339	0.201
Sample size	397	397	397	397

Note: Values in parentheses indicate t-value; * and ** represent the levels of significance at 10 % and 5 %, respectively

Table 7 *WTP* (truncated) estimation results

Category		PC model DFYES	 MBYES	PW model ASYM	 SYM
Truncated mean *WTP* (KRW per month, household)		KRW 1224.7 (US$ 0.98)	KRW 2333.4 (US$ 1.87)	KRW 2062.1 (US$ 1.65)	KRW 2747.5 (US$ 2.20)
95 % confidence interval [CI_L, CI_U]		[1013.8, 1435.6]	[2040.3, 2626.5]	[1802.4, 2321.8]	[2279.0, 3215.9]
Efficiency of *WTP* ((CI_U–CI_L)/mean *WTP*)		34.4 %	25.1 %	12.6 %	34.1 %
Yearly social benefits in the city* [CI_L, CI_U]		5.72 billion KRW [4.74, 6.71]	10.90 billions [9.53, 12.27]	8.03 billons [7.02, 9.04]	12.84 billions [10.65 ~ 15.02]
Total present value in the city	5 % discount rate [CI_L, CI_U]	120.14 billions [99.46, 140.85]	228.90 billions [200.13, 257.67]	168.63 billion [147.42, 189.84]	269.64 billions [223.65, 315.42]
	3 % discount rate [CI_L, CI_U]	196.42 billions [162.60, 230.27]	374.23 billions [327.20, 421.27]	275.69 billions [241.02, 310.37]	440.84 billions [365.65, 515.69]

Note: * is calculated based on 389,310 households of the Ulsan Metropolitan City in 2008

The 95 % confidence intervals in Table 7 are computed by the use of the Monte Carlo simulation technique suggested by Krinsky and Robb (1986, 1990), and the efficiency of WTP (EFWTP) is calculated by EFWTP = (CI_U–CI_L)/mean WTP, where CI_U and CI_L are upper and lower bounds of the 95 % confidence interval. The lower EFWTP indicates the higher efficiency of WTP estimate, because the lower

EFWTP means that the estimated WTP is more concentrated around the mean WTP. Comparing two models of DFYES and MBYES in the PC model, the MBYES model is more efficient in terms of EFWTP. This means that consideration of preference uncertainty results in the more efficient WTP estimation, as shown in Chang et al. (2007). Comparing two models of ASTM and SYM in the PW model, EFWTP is much lower in the ASYM model than the SYM model without statistically significant difference of the WTP between the two models. This indicates that the WTP estimation without consideration of preference uncertainty of the respondents with the negative willingness to pay results in the more efficient estimation. Comparing the two models of MBYES and ASYM, there is no statistically significant difference in terms of the WTP estimates. In terms of efficiency and goodness of fit (McFadden's R^2 in Table 6), however, the ASYM model provides better results than the MBYES model does. The MBYES model is similar to the dichotomous choice model with two choices of "yes" and "no." While both "definitely yes" and "maybe yes" are considered as the same "yes" in the MBYES model, the ASYM model provides "maybe yes" with the relatively lower weight than "definitely yes." In the choice reality, "definitely yes" and "maybe yes" are evidently different. Therefore, it is not surprising that the more efficient WTP is derived from the ASYM model than the MBYES model.

On the basis of the WTP estimate from the ASYM model with the most efficiency among the four models, the yearly social benefits from improvement and maintenance to swimming river water quality of the River Taehwa in the Ulsan Metropolitan City are estimated 8.03 billion KRW, and the total present value of the social benefits is estimated 275.69 billion KRW and 168.63 billion KRW at the 3.0 % and 5 % discount rate, respectively. The total investment of 133.0 billion KRW after 2008 is planned to improve and maintain the river water quality in the city. This represents that the total present value of the social benefits is greater than the total costs in terms of project investment by 2.1 times at the 3 % discount rate and by 1.3 times at the 5 % discount rate, respectively, in the ASYM model. Even in the DFYES model with the least WTP, the benefits are greater than the costs by 1.5 times at the 3 % discount rate, and the costs are included in the confidence interval of the benefit estimated at the 5 % discount rate. This result shows that the project to improve and maintain the river water quality of the downstream in the River Taehwa may be relevant in terms of cost and benefit. However, it is pointed out that this study has weakness in terms of cost/benefit analysis because the costs in this study do not consider the opportunity cost and other concomitant costs with the project.

6 Conclusion

This study analyzes the effects of the project investments on the river water quality improvement and also provides contingent valuation estimates of household's willingness to pay (WTP) for continuous public investment to the river water quality improvement and maintenance. The major findings from this study are as follows:

First, the estimation results using the OLS regression models with correction of autocorrelation show that the household soil pipe connection project with investment of 26.7 billion KRW reduced 1.68 ppm in BOD during the period of January 1998 to September 2007 and that the project dredging sediments at the river bottom with investment of 16 billion KRW resulted in the decrease of 1.12 ppm in BOD during the period of April 2004 to February 2007 at the downstream of the River Taehwa. The estimates of the monthly dummies show that the water quality during the water shortage period from December to June is significantly worse than that of October, but that of the period from July to November is not statistically different from that of October at the overall downstream. Although the monthly effects show the same pattern for all the four measurement points, the statistical significances are different among the four points. The monthly effects are getting weaker as the measure points are closer to the river mouth.

Second, based on a contingent valuation method with multiple choices in consideration of respondent's uncertainty, the estimation results of four logit models show that the probability of willingness to pay increases significantly with higher income, higher evaluation on river water quality, higher acknowledgement of the River Taehwa Master Plan, and higher age and decreases significantly with higher amount of bidding price. Truncated mean household's WTP is estimated from 1224.7 KRW to 2747 KRW for the respective four models. The present values of total social benefits in the Ulsan Metropolitan City are estimated from 196 billion KRW to 441 billion KRW for the respective four models, when applying the 3 % discount rate. This result shows that the present values of total social benefits are greater than the total costs in all models and thus may prove the economic relevancy of the project investments for the water quality improvement and maintenance in the city.

References

Alberini, A., K. Boyle, and M. Welsh. 2003. Analysis of contingent valuation data with multiple bids and response options allowing respondents to express uncertainty. *Journal of Environmental Economics and Management* 45: 40–62.

Arrow, R.J., E. Solow, E. Leamer, P. Portney, R. Radner, and H. Schuman. 1993. Report of the NOAA panel on contingent valuation. *Federal Register* 58: 4601–4614.

Berrens, R.P., H.K. Smith, A.K. Bohara, and C.L. Silva. 2002. Further investigation of voluntary contribution contingent valuation: Fair share, time contribution, and response uncertainty. *Journal of Environmental Economics and Management* 44: 144–168.

Cameron, T.A. 1988. A new paradigm for valuing non-market goods using referendum data: Maximum likelihood estimation by censored logistic regression. *Journal of Environmental Economics and Management* 15: 355–379.

Carson, R.T., W.M. Hanemann, R.J. Kopp, J.A. Krosnick, R.C. Mitchell, S. Presser, P.A. Ruud, and V.R. Smith. 1994. *Prospect interim lost use value due to DDT and PCB contamination in*

the Southern California bight. A Report of National Resource Damage Assessment, Inc. and Industrial Economics, Inc. to US NOAA.
Chang, J.I., S.H. Yoo, and S.J. Kwak. 2005. An empirical investigation of contingent valuation method with preference uncertainty. *Environmental and Resource Economics Review* 14(1): 75–245.
Chang, J.-I., S.-H. Yoo, and S.-J. Kwak. 2007. An investigation of preference uncertainty in the contingent valuation study. *Applied Economics Letters* 14: 691–695.
Greene, W. 2003. *Econometric Analysis*. 5th Edition, Prentice Hall, Englewood Cliffs.
Hanemann, W.M. 1984. Welfare evaluations in contingent valuation experiments with discrete response. *American Journal of Agricultural Economics* 66: 332–341.
Kim, J.H. 2007. Social benefits of improved water quality at the Taehwa river based on citizen's willingness-to pay. *Journal of Environmental Policy* 6(1): 83–110.
Kim, J.H. 2009. The effects of the project investment for water quality improvement in the river Taehwa. *Korea Spatial Planning Review* 62: 263–279.
Krinsky, I., and A.I. Robb. 1986. On approximating the statistical properties of elasticities. *Review of Economics and Statistics* 68: 713–719.
Krinsky, I., and A.I. Robb. 1990. On approximating the statistical properties of elasticities: Correction. *Review of Economics and Statistics* 72: 189–190.
Li, C.-Z., and L. Mattsson. 1995. Discrete choice under preference uncertainty: An improved structural model for contingent valuation. *Journal of Environmental Economics and Management* 28: 256–269.
Loomis, J., and E. Ekstrand. 1998. Alternative approaches for incorporating respondent uncertainty when estimating willingness to pay: The case of Mexican Spotted Owl. *Ecological Economics* 27: 29–42.
McConell, K.E. 1990. Models for referendum data: The structure of discrete choice models for contingent valuation. *Journal of Environmental Economics and Management* 18: 19–34.
Ready, R.C., J.C. Whitehead, and G.C. Bloomquist. 1995. Contingent valuation when respondents are ambivalent. *Journal of Environmental Economics and Management* 29: 181–196.
Wang, H. 1997. Treatment of "don't-know" responses in contingent valuation surveys. A random valuation model. *Journal of Environmental Economics and Management* 32: 219–232.
Welsh, M.P., and G.L. Poe. 1998. Elicitation effects in contingent valuation: Comparisons to a multiple bounded discrete choice approach. *Journal of Environmental Economics and Management* 38: 170–185.
Whitehead, J., G. Blomquist, R. Ready, and J. Huang. 1998. Construct validity of dichotomous choice contingent valuation questions. *Environmental and Resource Economics* 11: 107–116.
Whittington, D., G. Cassidy, D. Amaral, E. McClelland, H. Wang, and C. Poulos. 1994. The economic value of improving the environmental quality of Galveston Bay. The Calveston Bay national estuary program.
Yoo, S.H. 2007. Using one and one-half bounded dichotomous choice model to measure the economic benefit.

Trade and Environmental Responsibility for Greenhouse Gas Emissions: The Case of South Korea

Taelim Choi

Abstract A significant amount of embodied greenhouse gas (GHG) emissions have been and are currently being traded in the globalized economy. The conventional territorial approach to the control of GHGs released within a country fails to account for a large portion of GHGs for which a country may take responsibility, particularly from the perspective of consumption. Given the large volume of products traded among nations, a series of studies have underscored the need for the global monitoring of GHG emissions not only generated from production but also driven by consumptive activities. This study develops time-series GHG emission inventories from 1995 to 2009 from both production- and consumption-based perspectives in the case of South Korea and analyzes the factors that influence the increase and the decrease of GHG emissions. This empirical analysis has determined that production-based activities are more responsible for GHG emissions in South Korea than consumption-based activities. The analysis also found that the trade surplus of embodied GHG emissions in South Korea ranged from 0.31 to 1.01 tons per capita. A decomposition analysis showed that developments in environmental technology play a significant role in the reduction of GHG emissions, associated with a 45 % gross change in GHG emissions. However, this reduction was offset by increases in demand and changes in the input structure to energy-intensive sectors. The change of input structure is a critical factor contributing to trend in increasing embodied GHG emissions in not only South Korea but also nations linked with global trade.

Keywords Embodied emissions • Trade • Environmental responsibility • Structural decomposition

---

T. Choi (✉)
Incheon Development Institute, Incheon, Korea
e-mail: tlchoi@idi.re.kr

E. Kim, B.H.S. Kim (eds.), *Quantitative Regional Economic and Environmental Analysis for Sustainability in Korea*, New Frontiers in Regional Science: Asian Perspectives 25, DOI 10.1007/978-981-10-0300-4_13

1 Introduction

In the globalized economy, trade has had an increasingly significant impact on the environment. Therefore, as world economies become more closely linked, we must examine the association between trade and environmental changes from both local and global perspectives. The conventional approach for quantifying greenhouse gas (GHG) emissions, proposed by the Intergovernmental Panel on Climate Change (IPCC), is limited to national territories. The disadvantage of this territorial-based approach is that it overlooks the impact of environmental burden associated with trade. In fact, service-oriented countries passed their environmental burden to some developing countries depending on the energy-intensive industries from the consumption-based perspective. Through a trading relationship, they may accept less responsibility for emitting pollutants. As a complementary method of accounting for pollutants, the environmental multi-regional input-output (EMRIO) model captures GHG emissions beyond a national territory. Using the EMRIO model, we can measure the amount of embodied GHG emissions of internationally traded products and determine which countries are responsible for GHG emissions from both production and consumption activities.

Issues related to embodied emissions in trade and the responsibility for GHG emissions have been addressed in the literature. Several papers have proposed a conceptual approach to allocating GHG emissions from the perspectives of production or consumption (Eder and Narodoslawsky 1999; Rodrigues et al. 2006), a shared perspective of production and consumption (Lenzen et al. 2007), income-based responsibility (Marques et al. 2012), and downstream relationships (Lenzen and Murray 2010). These studies intended to clarify the attribution rule of various types of emissions. Some studies have conducted empirical analyses that quantified the consumption-based responsibility of the United Kingdom (Barrett et al. 2013) and Australian households (Lenzen and Peters 2010). Another has compared production- and consumption-based GHG inventories in the case of New Zealand (Andrew and Forgie 2008), and others have evaluated the trade balance of emissions for the Spanish economy (Serrano and Dietzenbacher 2010) and the Chinese economy (Pan et al. 2008) and in a case of bilateral trade between China and the United Kingdom (Li and Hewitt 2008) and the context of global trading systems (Peters et al. 2011). Finally, several have investigated IO-based GHG emissions inventories at the subnational scale in cases of US cities (Choi 2015), Wales (Turner et al. 2011), US households and cities (Jones and Kammen 2011), and the state of Oregon (Erickson, et al. 2012). These studies have indicated that an IO-based accounting method can play a complementary role in explaining how the extent of the responsibility for environmental burden of a country can vary according to the accounting principle.

The purpose of this study is to empirically develop time-series GHG emission inventories from both production and consumption perspectives by estimating the amount of embodied emissions from traded products in the case of South Korea. The findings of this analysis show a trend in the trade balance of embodied GHG emissions. In addition, from the results of a decomposition analysis, it attempts to explain the factors driving increases or decreases in the amount of embodied emissions of products. The case of South Korea provides useful insights into trends observed in the amounts of embodied emissions pertaining to a country, whose economy is heavily oriented to exporting and processing trade. This study utilizes the World Input-output Database (WIOD), which contains input-output tables and GHG emission statistics for 40 countries from 1995 to 2009 (Timmer et al. 2015). Using the data, it identifies a trend in embodied GHG emissions and causes for this trend stemming from changes in the global input-output structure and environmental technology.

The rest of this study consists of five sections. The next section discusses the principles of environmental responsibility from the viewpoint of geographical supply and demand of products. The third section describes a modeling framework used for estimating embodied GHG emissions in products, and the fourth section introduces the data source. The fifth section presents the results of the GHG emission inventories from three principles of accounting methods and the trade balance in the case of South Korea. The sixth section presents the results of the decomposition analysis and concludes with a discussion of future directions of research.

2 Trade and Environmental Responsibility

The total value of products traded worldwide has grown rapidly. Between 2005 and 2013, the annual growth rate of global trade was about 3.5 %, which was faster than that of world production, 2.0 %, and that of world GDP growth, 2.0 % (WTO 2014). Since a portion of demand met by imports has risen, the attribution of embodied emissions of traded products is key to our understanding of the extent to which a country is responsible for environmental emissions. The typology in Fig. 1 illustrates three types of emissions categorized by the geographical relationship of supply and demand as well as associated principles of environmental responsibility in a two-country context.

According to the geographical origins of the supply of and demand for products, in Box (A) of Fig. 1, GHG emissions can be characterized into three groups: Type 1, emissions that may be released as a result of domestic production activities and their products consumed within a country; Type 2, export-related emissions that may be also generated as a result of domestic production activities, but their products may also be meeting the demand of consumers in other countries; and Type 3, import-related emissions that may result from production activities outside of a country, but the products are meeting the demand of domestic consumers. The amount of environmental emissions for which a country takes responsibility differs according to how to allocate Type 2 and Type 3 emissions.

The various combinations of the three types of emissions yield three concepts of environmental responsibility for a country. Since the concept of production-based

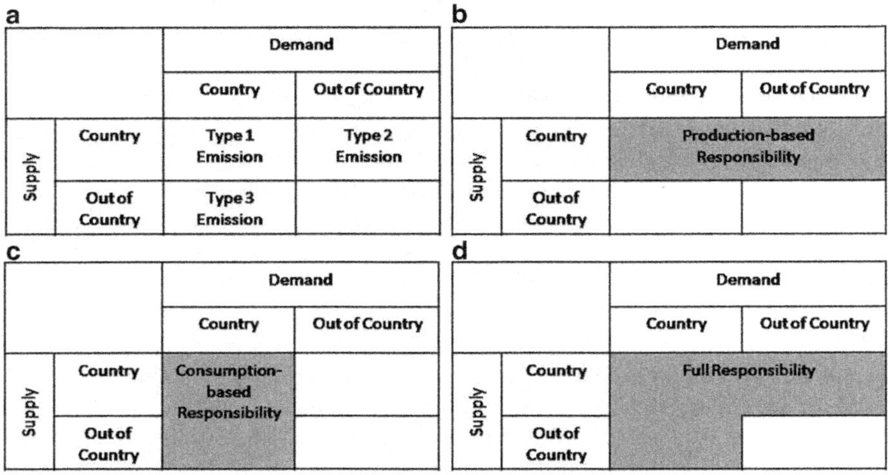

Fig. 1 Typology of environmental responsibility for country

responsibility includes emissions generated by a supply of products regardless of the geographic origin of demand, Types 1 and 2 emissions belong to production-based environmental responsibility. Such responsibility, therefore, excludes environmental burdens pertaining to the consumption of products supplied outside of a country, shown in Box (B) of Fig. 1. By contrast, consumption-based responsibility includes emissions originating from the consumption demand of a country regardless of geographical origin of supply of products. Hence, consumption-based responsibility of a country includes Types 1 and 3 shown in Box (C) in Fig. 1. Finally, we also assume that a country should take responsibility for any emissions released from all domestic productive and consumptive activities, including productive activities for exported products and the consumptive activities for imported products. Therefore, since a country could be responsible for all types of environmental emissions, this study refers to these emissions as the full responsibility of a country.

It should be noted that because this typology classifies emissions by geographical origins of the supply and demand of final products, not by the territory from which pollutants are actually released, each concept of responsibility may contain emissions released either within or beyond a territorial boundary. For example, Type 2 emissions consist of emissions of export-related products that may have some intermediate input imported from other countries. To calculate the embodied emissions in such export products, we have to consider upstream indirect emissions that are actually released from outside a country. As a result, any concept of environmental responsibility accounts for indirect emissions that may be released from either inside or outside of a country.

3 Modeling Framework for Environmental Responsibility

This section presents an environmental input-output modeling framework that elaborates on how Types 1, 2, and 3 emissions are estimated and allocated to a country with respect to three concepts of environmental responsibility. A conventional layout of the environmental input-output model with two countries, r and s, is presented in Table 1. Table 2 lists the explanations of every notation. Each country utilizes domestically supplied intermediate inputs ($\mathbf{z}^{\mathbf{rr}}$ and $\mathbf{z}^{\mathbf{ss}}$) and internationally supplied intermediate inputs ($\mathbf{z}^{\mathbf{rs}}$ and $\mathbf{z}^{\mathbf{sr}}$). The final demand consists of two components: the consumption of domestically supplied products ($\mathbf{f}^{\mathbf{rr}}$ and $\mathbf{f}^{\mathbf{ss}}$) and exports ($\mathbf{e}^{\mathbf{r}}$ and $\mathbf{e}^{\mathbf{s}}$). In the two-country case, the exports of country $r(\mathbf{e}^{\mathbf{r}})$ are equivalent to the consumption of imported products in country $s(\mathbf{f}^{\mathbf{rs}})$.

From the layout of the two-country context in Table 1, we present a modeling framework for quantifying the three types of emissions with respect to Box (A) of

Table 1 Layout of environmental two-region input-output table

		Industry		Final demand			
		Country r	Country s	Consumption	Export	Gross output	GHG emissions
Industry	Country r	$\mathbf{Z}^{\mathbf{rr}}$	$\mathbf{Z}^{\mathbf{rs}}$	$\mathbf{f}^{\mathbf{rr}}$	$\mathbf{e}^{\mathbf{r}}$ $(=\mathbf{f}^{\mathbf{rs}})$	$\mathbf{x}^{\mathbf{r}}$	$\mathbf{G}^{\mathbf{r}}$
	Country s	$\mathbf{Z}^{\mathbf{sr}}$	$\mathbf{Z}^{\mathbf{ss}}$	$\mathbf{f}^{\mathbf{ss}}$	$\mathbf{e}^{\mathbf{s}}$ $(=\mathbf{f}^{\mathbf{sr}})$	$\mathbf{x}^{\mathbf{s}}$	$\mathbf{G}^{s}$
Value added		$w^{\mathbf{r}}$	$w^{\mathbf{s}}$				
Gross output		$\mathbf{x}^{\mathbf{rT}}$	$\mathbf{x}^{\mathbf{sT}}$				

Table 2 Notation of variables in the environmental multi-region input-output model

Industry and country
Industry $= \{1, \cdots, n\}$, where industry consists of n number of conventional industries
r and s: each superscript denotes a country
GHG emissions
$\mathbf{G} = [G_i]$: greenhouse gas emissions by industry i
Type: a type of emissions, denoted by a subscript such as $\mathbf{G}_{\mathbf{Type1}}$, Type 1 emissions
g_i: greenhouse gas emissions by industry i per one dollar's worth of output of industry i
Variables
$\mathbf{Z} = [z_{ij}]$: intermediate input of industry i demanded by industry$j$
$\mathbf{x} = [x_i]$: gross output of industry i
$\mathbf{f} = [f_i]$: consumption for industry$i$
$\mathbf{e} = [e_i]$: export of industry$i$
$\mathbf{w} = [w_j]$: value added of industry j
$\mathbf{A} = [a_{ij}]$: intermediate input of industry$i$ purchased by industry j per one dollar's worth of output of industry j
$\mathbf{L} = [l_{ij}]$: total requirement of industry $i$purchased by industry j

Fig. 1. Type 1 emissions are associated with emissions generated from the consumption of domestically supplied products. Hence, the final demand term for Type 1 emissions is $\mathbf{f}^{\mathbf{rr}}$, whose emissions are determined by Eq. 1. Type 1 consists of direct and indirect emissions released from the production activities of country *r*, $\mathbf{G}^{\mathrm{r}}_{\mathbf{Type1}}$, and indirect emissions from the intermediate input supply from country *s*, $\mathbf{G}^{\mathrm{s}}_{\mathbf{Type1}}$:

$$\begin{bmatrix} \mathbf{G}^{r}_{\mathbf{Type1}} \\ \mathbf{G}^{s}_{\mathbf{Type1}} \end{bmatrix} = [g^{r} \quad g^{s}]\left(\mathbf{I} - \begin{bmatrix} \mathbf{A}^{\mathrm{rr}} & \mathbf{A}^{\mathrm{rs}} \\ \mathbf{A}^{\mathrm{sr}} & \mathbf{A}^{\mathrm{ss}} \end{bmatrix}\right)^{-1} \begin{bmatrix} \mathbf{f}^{\mathrm{rr}} \\ 0 \end{bmatrix} \tag{1}$$

Type 2 emissions are related to the production of exported products. The final demand term that drives Type 2 emissions is the export products of countryr, $\mathbf{e}^{\mathbf{r}}$. Type 2 emissions, determined by Eq. 2, consist of direct and indirect emissions released from the industrial activities of country *r* for export, $\mathbf{G}^{\mathrm{r}}_{\mathbf{Type2}}$, and indirect emissions from the intermediate input supply from countrys, $\mathbf{G}^{\mathrm{s}}_{\mathbf{Type2}}$:

$$\begin{bmatrix} G^{r}_{\mathbf{Type1}} \\ \mathbf{G}^{s}_{\mathbf{Type1}} \end{bmatrix} = [g^{r} \quad g^{s}]\left(\mathbf{I} - \begin{bmatrix} \mathbf{A}^{\mathrm{rr}} & \mathbf{A}^{\mathrm{rs}} \\ \mathbf{A}^{\mathrm{sr}} & \mathbf{A}^{\mathrm{ss}} \end{bmatrix}\right)^{-1} \begin{bmatrix} \mathbf{e}^{\mathrm{r}} \\ 0 \end{bmatrix} \tag{2}$$

Type 3 emissions originate from the final consumption of import products in country *r*. The final demand term represents imported products produced in country *s*, $\mathbf{f}^{\mathbf{sr}}$, and Type 3 emissions, determined by Eq. 3, include both the direct and indirect emissions released from region *s*, $\mathbf{G}^{\mathrm{s}}_{\mathbf{Type3}}$, and the indirect emissions from region *r*, $\mathbf{G}^{\mathrm{r}}_{\mathbf{Type3}}$:

$$\begin{bmatrix} \mathbf{G}^{r}_{\mathbf{Type3}} \\ \mathbf{G}^{s}_{\mathbf{Type3}} \end{bmatrix} = [g^{r} \quad g^{s}]\left(\mathbf{I} - \begin{bmatrix} \mathbf{A}^{\mathrm{rr}} & \mathbf{A}^{\mathrm{rs}} \\ \mathbf{A}^{\mathrm{sr}} & \mathbf{A}^{\mathrm{ss}} \end{bmatrix}\right)^{-1} \begin{bmatrix} 0 \\ \mathbf{f}^{\mathrm{sr}} \end{bmatrix} \tag{3}$$

Each concept of environmental responsibility in Boxes (B), (C), and (D) of Fig. 1 is comprised of combinations of the three types of emissions as follows: the production-based environmental responsibility of country *r* (including Types 1 and 2 emissions) is

$$\mathbf{G}^{r}_{\mathbf{Production}} = \mathbf{G}^{r}_{\mathbf{Type1}} + \mathbf{G}^{s}_{\mathbf{Type1}} + \mathbf{G}^{r}_{\mathbf{Type2}} + \mathbf{G}^{s}_{\mathbf{Type2}} \tag{4}$$

The consumption-based environmental responsibility of countryr (including Types 1 and 3 emissions) is

$$\mathbf{G}^{r}_{\mathbf{Consumption}} = \mathbf{G}^{r}_{\mathbf{Type1}} + \mathbf{G}^{s}_{\mathbf{Type1}} + \mathbf{G}^{r}_{\mathbf{Type3}} + \mathbf{G}^{s}_{\mathbf{Type3}} \tag{5}$$

The full environmental responsibility of countryr (including Types 1, 2, and 3 emissions) is

$$\mathbf{G}^{r}_{\mathbf{Full}} = \mathbf{G}^{r}_{\mathbf{Type1}} + \mathbf{G}^{s}_{\mathbf{Type1}} + \mathbf{G}^{r}_{\mathbf{Type2}} + \mathbf{G}^{s}_{\mathbf{Type2}} + \mathbf{G}^{r}_{\mathbf{Type3}} + \mathbf{G}^{s}_{\mathbf{Type3}} \tag{6}$$

The trade balance of embodied emissions is the difference between the amount of GHG emissions of production-based responsibility and that of consumption-based responsibility

$$\mathbf{G}^{r}_{\textbf{Trade Balance}} = \mathbf{G}^{r}_{\textbf{Production}} - \mathbf{G}^{r}_{\textbf{Consumption}} \tag{7}$$

4 Data

The World Input-output Database (WIOD) provides international multi-region input-output data. It covers 16 years from 1995 to 2009, 40 countries with 35 industry sectors from developed countries such as those of the European Union and the United States and developing countries such as China, Indonesia, India, Brazil, and South Korea. It contains information about the origin and the destination of intermediate and final products by country and industry as well as sectoral GHG emissions used for measuring the intensity of the emissions. The WIOD includes the main trading partners of South Korea: China, Japan, and the United States. The data are useful for estimating the embodied emissions of South Korean products as well as foreign products imported to South Korea for consumption. Hence, with the WIOD, we can estimate time-series embodied GHG emission inventories for South Korea and associate them with the concepts of environmental responsibility.

5 Results

This study develops time-series GHG emissions inventories for South Korea by quantifying embodied emissions driven by three types of final demand consumption. Figure 2 presents per capita GHG emissions resulting from (1) the South Korean consumption of domestically supplied products, (2) exports of South Korean products, and (3) the South Korean consumption of imported products. It shows the portion that international trade contributes to the GHG emissions of South Korea, which amounted to about 24.1 % in 2009. In 2009, per capita GHG emissions from the exports of South Korean products were 1.89 tons (16.5 %) and those of South Korean consumption of imported products were 0.88 (7.6 %) tons. Per capita GHG emissions generated from the consumption of domestically supplied products entailed 75.9 % of South Korean GHG emissions. In addition, the portion of GHG emissions from international trade has gradually increased over time. The percentage of GHG emissions from exports and imports was 18.1 % in 1995, 22.9 % in 2000, 23.5 % in 2005, and 24.5 % in 2009. Comparing 1995 and 2009, the portion of GHG emissions from exports rose by 6.0%p and that from consumption of imported products by 0.7%p. The export of final products produced

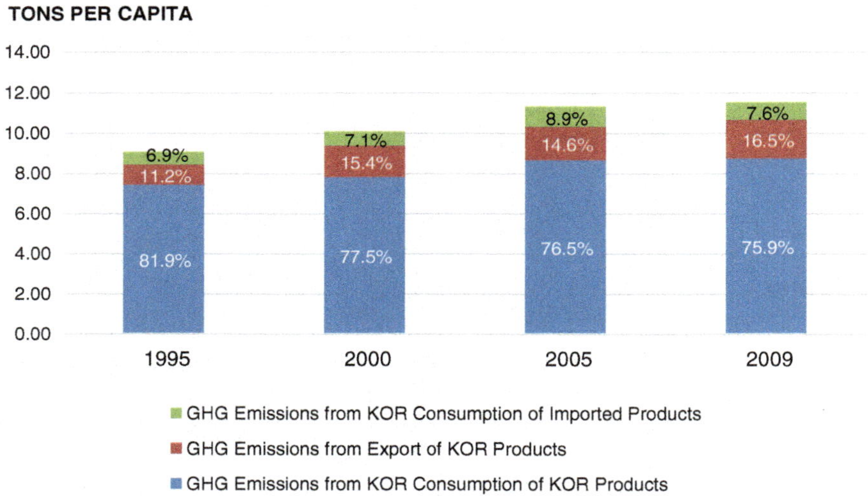

Fig. 2 GHG emissions by types of final demand

in South Korea played a significant role in the increase in the GHG emissions of South Korea.

From the estimates of embodied GHG emissions related to the three types of final demands, GHG emission inventories regarding product-based, consumption-based, and full responsibility are shown in Fig. 3. In 2009, per capita GHG emissions related to production-based responsibility in South Korea were 10.62 tons while those of consumption-based responsibility were 9.60 tons. That is, the amount of GHG emissions related to production-based responsibility was slightly higher, 1.01 tons per capita, than that related to consumption-based responsibility. These relationships were consistent in 1995, 2000, and 2005. GHG emissions associated with full responsibility were 11.5 tons per capita in 2009.

The higher level of responsibility for production-based GHG emissions than that for consumption-based emissions indicates that the South Korean economy had a surplus in the trade margin of GHG emissions from 1995 and 2009. The trade balance of GHG emissions is calculated by subtracting the amount of the GHG emissions of imports from the amount from exports. During each year, GHG emissions released from exported Korean products outweighed those from the consumption of imported products. The trade surplus of GHG emissions varied between 0.31 and 1.01. However, it increased during periods of economic crisis in South Korea. Indeed, during the Asian financial crisis of 1998, the surplus of GHG emissions rose to 0.99 tons per capita, and during the global economic crisis of 2009, it increased to 1.01 tons per capita. These increases may have been due to the more severe impact they had on household expenditures than on the competitiveness of the export-based industries of South Korea. This inconsistent impact led to wider gaps between the GHG emissions of embodied emissions from exports and those from imports in South Korea (Fig. 4).

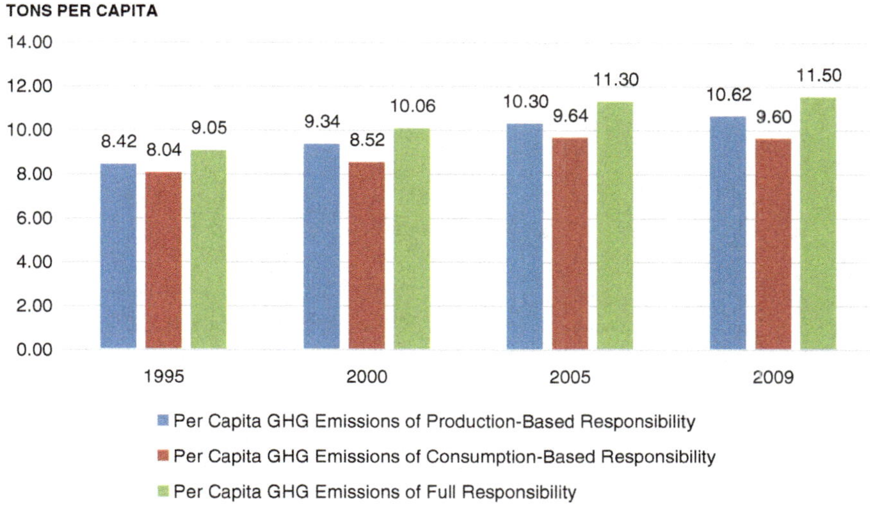

Fig. 3 Environmental responsibility of the South Korean economy

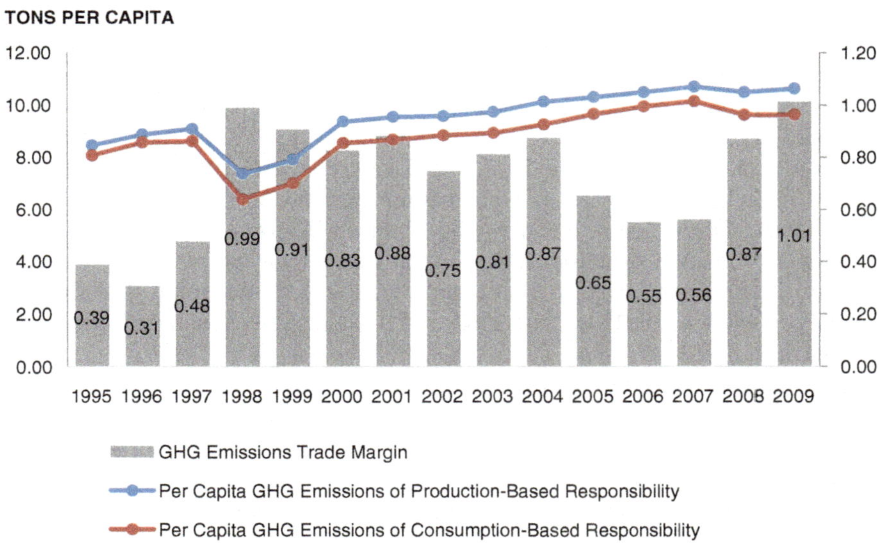

Fig. 4 Environmental responsibility and GHG emission trade margin

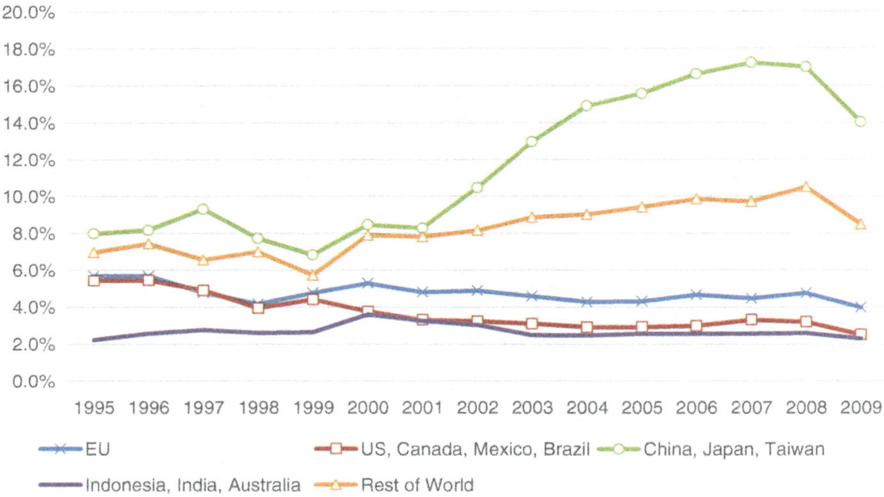

Fig. 5 GHG emissions by country

The EMRIO analysis also displays the amount of GHG emissions by geographical area. In 1995, the portion of GHG emissions released within South Korea was 71.9 %, but in the 2000s, it started to decline: 68.1%in 2003, 65.3 % in 2005, and 62.1 % in 2008. The decline of GHGs emitted in the territorial boundary of South Korea was largely associated with the increased portion of GHGs emitted in the territorial boundary of China, shown in Fig. 5, which sharply increased after 2001. The portion of GHG emissions released in China was 5.9 % in 2001 and rose to 13.5 % in 2007. From 2001 to 2007, the portion of GHG emissions released in South Korea declined by 9.8 %p while that released in China increased by 7.6 %p. By contrast, the portion released by the European Union and the United States continued to decline over the same period. The percentage of GHG emissions released in both the European Union and the United States dropped slightly: in the former, 4.8 % in 2001, 4.4 % in 2007, and 3.9 % in 2009, and in the latter, 2.6 % in 2001, 2.7 % in 2007, and 1.9 % in 2009.

The manufacturing, utility, and transportation sectors were more than 80 % of GHG emissions in the concept of full responsibility of South Korea in 2000s. The sectoral compositions have been relatively stable over time except for a decline in the manufacturing sector and an increase in the utility sector. In 1995, manufacturing was the largest contributor to the GHG emissions of South Korea, over 40 %; however, this percentage decreased to 38.1 % in 2000, 32.6 % in 2005, and 30.0 % in 2009. By contrast, utilities increased markedly from 25.1 % in 1995 to 39.5 % in 2005 and 44.0 % in 2009 (Fig. 6).

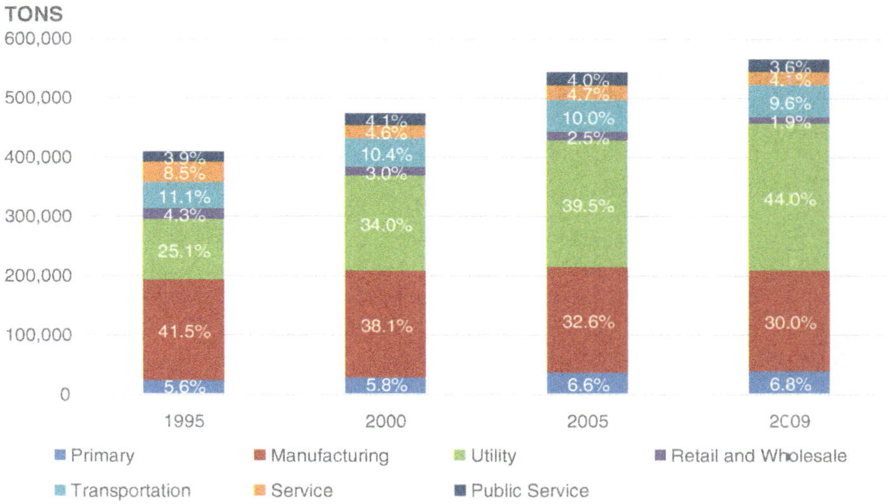

Fig. 6 GHG emissions by industrial sector with full responsibility in South Korea

6 Decomposition of Change in the GHG Emissions of South Korea

The EMRIO analysis indicated that the amount of GHG emissions for which South Korea should take responsibility increased from the perspectives of production and consumption. Using a decomposition analysis, this section examines the factors that drive changes in the amount of GHG emissions in full responsibility. The decomposition analysis is a conventional technique used for decomposing observed changes into several components (Miller and Blair 2009). In this study, changes in the amount of GHG emissions are decomposed into three types of changes: (1) environmental technology, (2) final demand, and (3) input structure. A change in GHG emissions ($\mathbf{\Delta G_{(t,\ t+n)}}$) during two time periods ($\mathbf{t},\ \mathbf{t+n}$) is expressed as

$$\mathbf{\Delta G_{(t,\ t+n)} = G_{t+n} - G_t = g_{t+n}x_{t+n} - g_t x_t} \tag{8}$$

Equation 8 can be rearranged into the two forms expressed in Eqs. 9 and 10:

$$\begin{aligned}\mathbf{\Delta G_{(t,\ t+n)}} &= \mathbf{g_{t+n}(x_{t+n} - x_t) + (g_{t+n} - g_t)x_t} \\ &= \mathbf{g_{t+n}\Delta x_{(t,\ t+n)} + x_t \Delta g_{(t,\ t+n)}}\end{aligned} \tag{9}$$

$$\begin{aligned}\mathbf{\Delta G_{(t,\ t+n)}} &= \mathbf{(g_{t+n} - g_t)x_{t+n} + g_t(x_{t+n} - x_t)} \\ &= \mathbf{\Delta g_{(t,t+n)}x_{\ t+n} + g\Delta x_{(t,\ t+n)}}\end{aligned} \tag{10}$$

From an average of Eqs. 9 and 10, the changes in the amount of GHG emissions are decomposed into two components, as shown in Eq. 11. The first component of Eq. 11 is the effect of environmental technological change, and the second is the effect of output change:

$$\Delta \mathbf{G}_{(t,\ t+n)} = \underbrace{\frac{1}{2}\Delta \mathbf{g}_{(t,\ t+n)}(\mathbf{x}_t + \mathbf{x}_{t+n})}_{\text{Environmental Technology Change}} + \underbrace{\frac{1}{2}\Delta \mathbf{x}_{(t,\ t+n)}(\mathbf{g}_t + \mathbf{g}_{t+n})}_{\text{Output Change}} \tag{11}$$

Following the same steps, the output change is also decomposed into two components of the final demand change and the input structure change. The output change is expressed as Eq. 12:

$$\Delta \mathbf{x}_{(t,t+n)} = \mathbf{x}_{t+n} - \mathbf{x}_t = \mathbf{L}_{t+n}\mathbf{f}_{t+n} - \mathbf{L}_t\mathbf{f}_t \tag{12}$$

Equation 12 is rearranged into the two forms of Eqs. 13 and 14:

$$\begin{aligned} \Delta \mathbf{x}_{(t,t+n)} &= \mathbf{L}_{t+n}(\mathbf{f}_{t+n} - \mathbf{f}_t) + (\mathbf{L}_{t+n} - \mathbf{L}_t)\mathbf{f}_t \\ &= \mathbf{L}_{t+n}\Delta \mathbf{f}_{(t,\ t+n)} + \mathbf{f}_t\Delta \mathbf{L}_{(t,\ t+n)} \end{aligned} \tag{13}$$

$$\begin{aligned} \Delta \mathbf{x}_{(t,t+n)} &= (\mathbf{L}_{t+n} - \mathbf{L}_t)\mathbf{f}_{t+n} + \mathbf{L}_t(\mathbf{f}_{t+n} - \mathbf{f}_t) \\ &= \Delta \mathbf{L}_{(t,t+n)}\mathbf{f}_{t+n} + \mathbf{L}_t\Delta \mathbf{f}_{(t,\ t+n)} \end{aligned} \tag{14}$$

As expressed in Eq. 15, the output change is comprised of the first component of the final demand change and the second component of the input structure change by taking an average of Eqs. 13 and 14:

$$\Delta \mathbf{x}_{(t,t+n)} = \underbrace{\frac{1}{2}\Delta \mathbf{f}_{(t,\ t+n)}(\mathbf{L}_t + \mathbf{L}_{t+n})}_{\text{Final Demand Chage}} + \underbrace{\frac{1}{2}\Delta \mathbf{L}_{(t,\ t+n)}(\mathbf{f}_t + \mathbf{f}_{t+n})}_{\text{Input Structure Change}} \tag{15}$$

When the decomposed components of the output change are combined into Eq. 11, the change in the amount of GHG emissions is explained by the three components—changes in environmental technology change, final demand, and the input structure—expressed in Eq. 16:

$$\begin{aligned} \Delta \mathbf{G}_{(t,\ t+n)} = &\underbrace{\frac{1}{2}\Delta \mathbf{g}_{(t,\ t+n)}(\mathbf{x}_t + \mathbf{x}_{t+n})}_{\text{Environmental technological Change}} + \underbrace{\frac{1}{4}\Delta \mathbf{f}_{(t,\ t+n)}(\mathbf{L}_t + \mathbf{L}_{t+n})(\mathbf{g}_t + \mathbf{g}_{t+n})}_{\text{Final Demand Change}} \\ &+ \underbrace{\frac{1}{4}\Delta \mathbf{L}(\mathbf{f}_t + \mathbf{f}_{t+n})(\mathbf{g}_t + \mathbf{g}_{t+n})}_{\text{Input Structure Change}} \end{aligned} \tag{16}$$

This decomposition technique explains changes in the amount of GHG emissions between 2002 and 2007, which excludes the economic recessions. The results show a net change of 82,740 thousand tons during the 5-year period and a gross change of 804,873 thousand tons, the absolute sum of the three components. The gross change consists of changes resulting from environmental technology, 44.9 %; final demand, 42.0 %; and input structure, 13.1 %. Overall, the effects of changes (increased amounts of GHG emissions) in the final demand and the input structure outweighed those (reduced amounts) of environmental technology, for the latter did not lead to an overall reduction in GHG emission level.

Nevertheless, the effect of such advances in environmental technology appears to be significant. The reduced amount of GHG emissions originating from environmental technological advances (361,066 tons) exceeded the increased amount of GHG emissions originating from increased final demand (337,979 tons). Changes in the input structure exerted an additional environmental burden of 105,827 tons of GHG emissions. Although the magnitude of the effect of changes in the input structure, compared to other components, was relatively small, it strongly contributes to the increase in embodied GHG emissions in South Korea (Fig. 7).

The three decomposed effects are rearranged by country in Table 3 and by industry in Table 4, which clarified some key patterns about net GHG gains of South Korea. The first pattern shows that the net gain of GHG emissions released in a territory of South Korea comprises only 18 % of the net changes in GHG emissions. While the reduction resulting from changes in environmental technology in South Korea was 225,142 thousand tons, the amount of GHG emissions resulting from changes in the final demand was 213,609 thousand tons. The environmental technology of South Korea appears to advance rapidly enough to offset the increase in GHG emissions resulting from increases in the final demand from 2002 to 2007. In the territorial boundary of South Korea, changes in the input structure (26,619 thousand tons) had a relatively minor impact on the amount of released GHG emissions

Another pattern of net GHG gains in the emissions of South Korea is that more than 48 % of the net increase in emissions (40,106 tons) can be attributed to China, the largest contributor to the net gain in GHG emissions, shown in Table 3. In particular, the component of changes in the input structure of China substantially

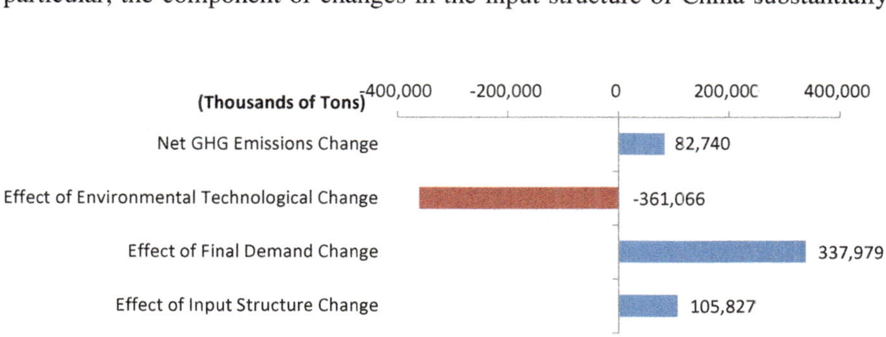

Fig. 7 Structural decomposition of changes in GHG emissions in South Korea (2002–2007)

Table 3 Decomposition of GHG emissions by country

GHG emissions change (in thousands of tons)				
Country	Effect of environmental technological change	Effect of final demand change	Effect of input structure change	Net GHG emissions change
Australia	−4657	3739	939	21
China	−48,109	44,431	43,784	40,106
EU	−9711	8730	3602	2621
Indonesia	−3149	2934	−773	−989
Japan	−4728	8028	2088	5388
Russia	−19,373	9953	8372	−1048
United States	−4987	7910	−52	2871
South Korea	−225,142	213,609	26,619	15,086
Taiwan	−1234	2287	1030	2083
Row	−39,977	36,358	20,220	16,601
Total	−361,066	337,979	105,827	82,740
Percentage by countries				
Australia	1.3 %	1.1 %	0.9 %	0.0 %
China	13.3 %	13.1 %	41.4 %	48.5 %
EU	2.7 %	2.6 %	3.4 %	3.2 %
Indonesia	0.9 %	0.9 %	−0.7 %	−1.2 %
Japan	1.3 %	2.4 %	2.0 %	6.5 %
Russia	5.4 %	2.9 %	7.9 %	−1.3 %
United States	1.4 %	2.3 %	0.0 %	3.5 %
South Korea	62.4 %	63.2 %	25.2 %	18.2 %
Taiwan	0.3 %	0.7 %	1.0 %	2.5 %
Row	11.1 %	10.8 %	19.1 %	20.1 %
Total	100.0 %	100.0 %	100.0 %	100.0 %

affects net GHG emissions by adding 43,784 thousand tons, suggesting that the transfer of Korean supply chain to China heavily affected the level of embodied GHG emissions in South Korea. Third, the utility sector accounts for almost 80 % of changes in net GHG emissions (65,957 thousand tons). Changes in the input structure were a key element of net increase of utility sector (60,839 thousand tons), indicating that the input supply from the utility sector increased. This pattern of change was associated with the transfer of the Korean supply chain to China. The amount of inputs from the Chinese utility sector substantially rose between 2002 and 2007.

Table 4 Decomposition of GHG emissions by industry

GHG emission change (in thousands of tons)				
Country	Effect of environmental technological change	Effect of final demand change	Effect of input structure change	Net GHG emissions change
Primary	−27,851	21,017	6685	−149
Refinery and chemicals	−35,040	29,341	12,001	6302
Nonmetallic minerals	−15,123	26,066	−7092	3851
Basic and fabricated metals	−64,828	47,294	25,469	7934
Other manufacturing	−23,424	18,133	589	−4702
Utilities	−120,667	125,784	60,839	65,957
Retail and wholesale	−9720	7000	−182	−2901
Transportation	−36,267	33,689	7282	4704
Services	−13,408	14,678	−150	1119
Public services	−14,739	14,977	388	626
Total	−361,066	337,979	105,827	82,740
Percentage by industry				
Primary	7.7 %	6.2 %	6.3 %	−0.2 %
Refinery and chemicals	9.7 %	8.7 %	11.3 %	7.6 %
Nonmetallic minerals	4.2 %	7.7 %	−6.7 %	4.7 %
Basic and fabricated metals	18.0 %	14.0 %	24.1 %	9.6 %
Other manufacturing	6.5 %	5.4 %	0.6 %	−5.7 %
Utilities	33.4 %	37.2 %	57.5 %	79.7 %
Retail and wholesale	2.7 %	2.1 %	−0.2 %	−3.5 %
Transportation	10.0 %	10.0 %	6.9 %	5.7 %
Services	3.7 %	4.3 %	−0.1 %	1.4 %
Public services	4.1 %	4.4 %	0.4 %	0.8 %
Total	100.0 %	100.0 %	100.0 %	100.0 %

7 Conclusion

This study empirically analyzed greenhouse gas (GHG) emissions inventories with respect to various concepts of environmental responsibility in the case of South Korea. From the environmental multi-regional input-output (EMRIO) model, this study found that the per capita GHG emissions of South Korea increased from both the perspectives of production and consumption and that South Korea exhibited a surplus in the trade of embodied emissions associated with the trade of products. However, the increased net volume of embodied GHG emissions was attributed to input structure change. Much of effect of input structure change occurred in China. It indicated that the effect of input structure change is associated with the transfer of much of the supply chain to more energy-intensive sector sand countries such as China. As a result, the impact on the net gain in GHG emissions in the territorial boundary of South Korea is relatively small, while the intensification of the Korean supply chain in China significantly impacted the amount of embodied GHG emissions related to the production and consumption of South Korea. Hence, this study indicates that international supply chain has significant impact on GHG emissions and choices in demand of products would be very relevant to action plans to reduction of GHG emissions. The GHG reduction policies of South Korea mostly focus on advance of technological control and introduction of cap-and-trade market. While consumption issues are rarely addressed as a policy option, cooperation among entities in the supply chain and consumption-side policy may contribute to mitigating global environmental burdens.

References

Andrew, R., and V. Forgie. 2008. A three-perspective view of greenhouse gas emission responsibilities in New Zealand. *Ecological Economics* 68(1–2): 194–204.

Barrett, J., G. Peters, T. Wiedmann, K. Scott, M. Lenzen, K. Roelich, and C. Le Quéré. 2013. Consumption-based GHG emission accounting: A UK case study. *Climate Policy* 13(4): 451–470.

Choi, T. 2015. Understanding environmental responsibility of cities and emissions embodied in trade. *Economic Systems Research* 27(2): 133–153.

Eder, P., and M. Narodoslawsky. 1999. What environmental pressures are a region's industries responsible for? A method of analysis with descriptive indices and input–output models. *Ecological Economics* 29(3): 359–374.

Erickson, P., D. Allaway, M. Lazarus, and E.A. Stanton. 2012. A consumption-based GHG inventory for the U.S. state of Oregon. *Environmental Science & Technology* 46(7): 3679–3686.

Jones, C.M., and D.M. Kammen. 2011. Quantifying carbon footprint reduction opportunities for U.S. households and communities. *Environmental Science & Technology* 45(9): 4088–4095.

Lenzen, M., and J. Murray. 2010. Conceptualising environmental responsibility. *Ecological Economics* 70(2): 261–270.

Lenzen, M., and G.M. Peters. 2010. How city dwellers affect their resource hinterland. *Journal of Industrial Ecology* 14(1): 73–90.

Lenzen, M., J. Murray, F. Sack, and T. Wiedmann. 2007. Shared producer and consumer responsibility — Theory and practice. *Ecological Economics* 61(1): 27–42.

Li, Y., and C.N. Hewitt. 2008. The effect of trade between China and the UK on national and global carbon dioxide emissions. *Energy Policy* 36(6): 1907–1914.

Marques, A., J. Rodrigues, M. Lenzen, and T. Domingos. 2012. Income-based environmental responsibility. *Ecological Economics* 84: 57–65.

Miller, R.E., and P.D. Blair. 2009. *Input-output analysis: Foundations and extensions*, 2nd ed. Cambridge: Cambridge University Press.

Pan, J., J. Phillips, and Y. Chen. 2008. China's balance of emissions embodied in trade: Approaches to measurement and allocating international responsibility. *Oxford Review of Economic Policy* 24(2): 354–376.

Peters, G.P., J.C. Minx, C.L. Weber, and O. Edenhofer. 2011. Growth in emission transfers via international trade from 1990 to 2008. *Proceedings of the National Academy of Sciences* 108 (21): 8903–8908.

Rodrigues, J., T. Domingos, S. Giljum, and F. Schneider. 2006. Designing an indicator of environmental responsibility. *Ecological Economics* 59(3): 256–266.

Serrano, M., and E. Dietzenbacher. 2010. Responsibility and trade emission balances: An evaluation of approaches. *Ecological Economics* 69(11): 2224–2232.

Timmer, M.P., E. Dietzenbacher, B. Los, R. Stehrer, and G.J. Vries. 2015. An illustrated user guide to the world input–output database: The case of global automotive production. *Review of International Economics*. 23: 575–605.

Turner, K., M. Munday, S. McIntyre, and C.D. Jensen. 2011. Incorporating jurisdiction issues into regional carbon accounts under production and consumption accounting principles. *Environment and Planning A* 43(3): 722–741.

World Trade Organization. 2014. International Trade Statistics 2014. https://www.wto.org/english/res_e/statis_e/its2014_e/its14_toc_e.htm. Accessed 10 Aug 2015.

CPSIA information can be obtained
at www.ICGtesting.com
Printed in the USA
LVOW02*1809221116
514102LV00002B/47/P

9 789811 002984